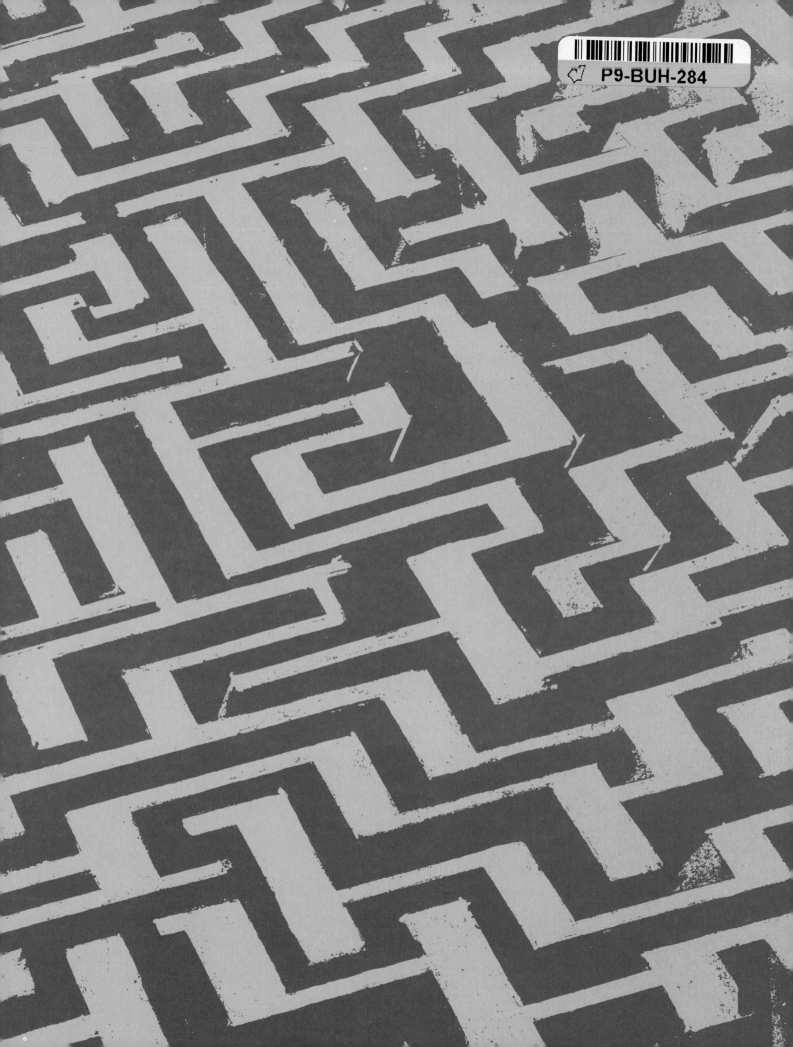

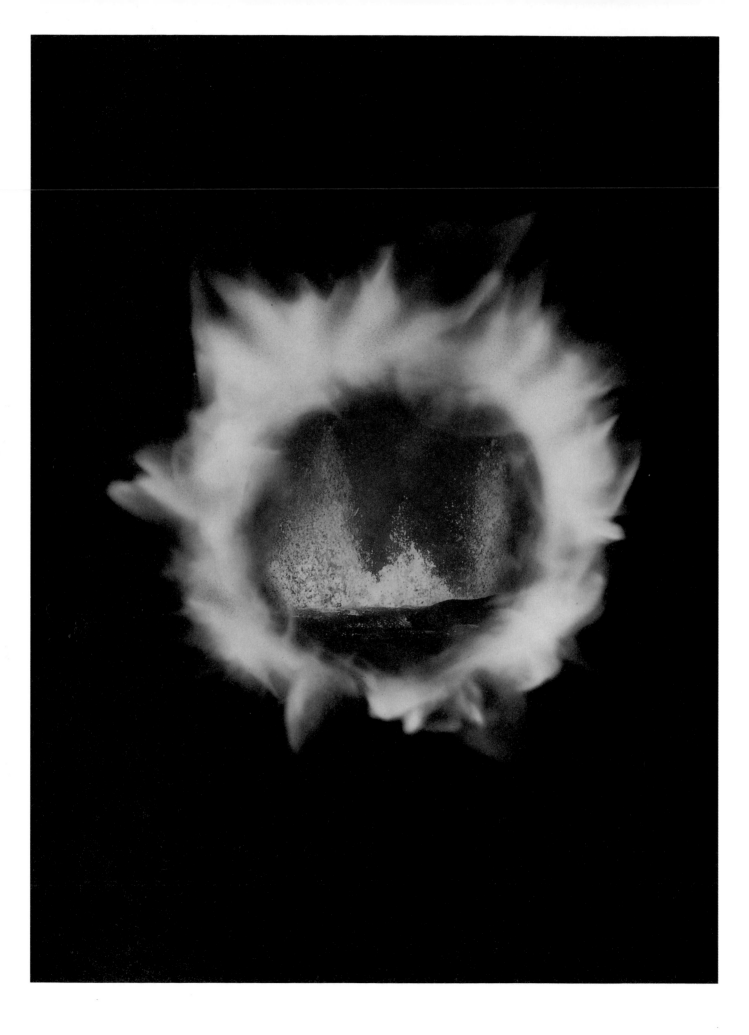

QUEST FOR THE UNKNOWN

SECRETS
OF THE
NATURAL WORLD

THE READER'S DIGEST ASSOCIATION, INC.
Pleasantville, New York/Montreal

Quest for the Unknown
Created, edited, and designed by DK Direct Limited

A Dorling Kindersley Book

DK Direct Limited

Series Editor Richard Williams
Senior Editor Ellen Dupont; **Editors** Kim Inglis, Maxine Lewis
Editorial Research Julie Whitaker

Senior Art Editor Susie Breen
Art Editor Mark Batley; **Designer** Sara Hill
Senior Picture Researcher Frances Vargo; **Picture Researcher** Lesley Coleman
Picture Assistant Sharon Southren

Editorial Director Jonathan Reed; **Design Director** Ed Day
Production Manager Ian Paton

Volume Consultant Dr. Karl P. N. Shuker
Commissioning Editor Peter Brookesmith
Contributors Dennis Bardens, Michael Bright, Peter Brookesmith, Prof. J. L. Cloudsley-Thompson,
Nicholas Jones, William Lindsay, Dr. Malcolm MacGarvin, Lesley Riley, Dr. Karl P. N. Shuker

Illustrators Roy Flooks, Andrew Griffin, Tracey Hughes, Janos Marffy, Gary Marsh,
Steve Rawlings, Darrel Rees, Matthew Richardson, Indra Sharma, Steve Wallace
Photographers Mark Hamilton, Alex Wilson
Modelmakers Atlas Models

Library of Congress Cataloging in Publication Data

Secrets of the natural world.
    p. cm. — (Quest for the unknown)
"A Dorling Kindersley book" — T.p. verso.
Includes index.
ISBN 0-89577-498-4
1. Natural history—Popular works. 2. Natural history—
Miscellanea. I. Reader's Digest Association. II. Series.
QH45.5.S424 1993
508—dc20   93-435

Printed in the United States of America

FOREWORD

*F*ROM TIME IMMEMORIAL, THE EARTH has harbored secrets that humans have been unable to explain. In spite of the recent advances in botany, biology, zoology, and geology, we still do not fully understand many of the mysteries of plant and animal life, or of the physical geography of the planet itself.

In this volume, we examine amazing animal abilities: how some creatures possess the power to withstand huge doses of radiation, how others are able to undertake extraordinary migratory journeys, and how yet others have developed "super senses." We present such controversial anomalies as rat kings, luminous owls, leaping snakes, and animals that seem to have hypnotic powers.

The plant kingdom is here revealed in all its extraordinary diversity. *Secrets of the Natural World* shows how some plants have the power to heal, and others to harm; how some can shine a beam of light, and others have become carnivores. And the book weighs the evidence that some plants are apparently able to communicate with others of their own kind.

Perhaps the ultimate mystery lies under our feet, in the makeup of our ever-changing planet. In the ongoing quest for understanding, you'll find in these pages an exploration of such phenomena as singing sands and ringing rocks, earthquake lights and shining seas — and the delicate and continuing balance that is needed to sustain all life on earth.

— The Editors

CONTENTS

FOREWORD
FOREWORD
5

INTRODUCTION
SAVED BY DOLPHINS
8

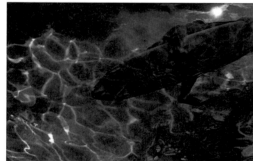

CHAPTER ONE
SECRET ANIMALS
18

ANIMAL HYPNOTISTS
SERPENT RIDDLES
UNNATURAL ANIMALS
RAT KINGS
SURVIVORS FROM THE PAST?
FROM FOLKLORE TO FACT

CHAPTER TWO
MYSTERIOUS SENSES
42

SECRET SCENTS
ULTRA-VISION
BEASTLY FORECASTS
EXTRAORDINARY ANIMAL POWERS
CASEBOOK
ANIMAL LANGUAGE

CHAPTER THREE
SECRETS OF SURVIVAL
62

THRIVING ON POISON
ANIMAL PLAGUES
INVADERS FROM AFAR
INEXPLICABLE INTERACTIONS
RUNNING WILD
TEMPERATURE EXTREMES

CHAPTER FIVE
THE WORLD OF PLANTS
104

FABULOUS FUNGI
PHARMACOPEIA OF THE FUTURE
PLANTS THAT EAT FLESH
MYSTERIES OF PLANT BEHAVIOR
DECEIT AND DECEPTION
HERBAL TRADITIONS: PLANTS THAT HEAL
PLANTS THAT HARM

CHAPTER FOUR
ENIGMAS OF EXTINCTION
78

THE BURGESS SHALE
DEATH OF THE DINOSAURS
DINOSAURS AT THE MOVIES
DEAD ENDS
DINOSAUR DISCOVERIES
MAMMOTHS
LIVING FOSSILS
REVERSING EXTINCTION
IS THE DODO REALLY DEAD?

CHAPTER SIX
WONDERS OF THE PLANET
124

GENUINE SPARKLERS
VOYAGE TO THE SECRET VENT WORLD
MYSTERIES OF THE DEEP
CASEBOOK
THE GAIA HYPOTHESIS

INDEX
142

SAVED BY DOLPHINS

There are many reports of dolphins rescuing the victims of shark attacks and shipwrecks, and even guiding beached whales to safety. Yet the mystery remains: In the ultracompetitive natural world, why would they act unselfishly?

Three Australian teenagers once had an unforgettable encounter with a school of dolphins. The boys — Adam Maguire, Jason Moloney, and Bradley Thompson — all 17, came from Ballina in New South Wales, 100 miles south of Brisbane. During their summer holidays, in January 1989, they decided to go surfing off isolated Halftide Beach at nearby Evans Head.

It was a perfect day, with bright summer sun and crashing surf. Then a school of about 20 dolphins joined the boys and started frolicking with them in the waves.

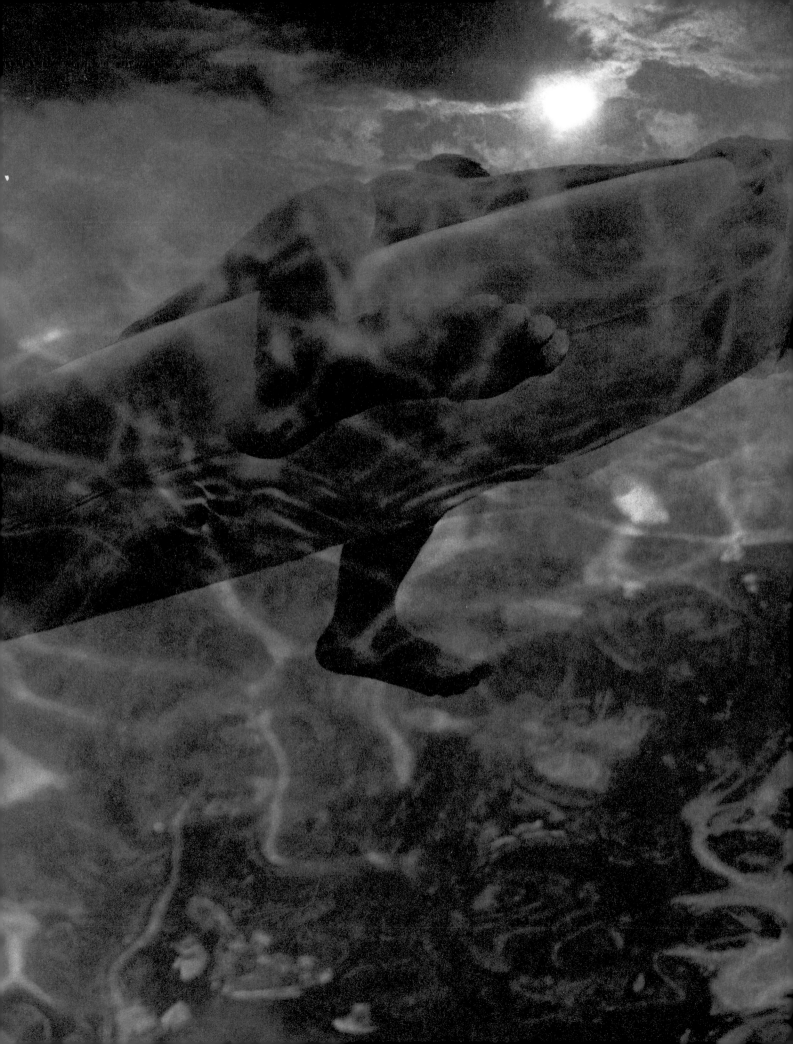

After an hour, the dolphins' behavior changed. They became restless, and instead of swimming in to the beach with the boys, they formed a circle around the surfers and swam not only around them but also below them. At the time, the boys assumed this new kind of activity was simply another form of dolphin play. But later, the boys came to believe that the dolphins might have been trying to give them warning of, or possibly even protection from, a deadly marauder that was lurking in the sea.

Moving in for the kill

According to Adam Maguire, the school of dolphins began to jump frantically and make a lot of noise: "Then I saw what looked like a larger dolphin. I didn't take much notice until it was a few yards away. Then I saw its fin and realized it was a shark." The shark, which the boys estimated was between 10 and 13 feet long, surfaced beneath Adam's surfboard just as he caught a wave and was riding in toward the beach. It ripped a huge chunk out of the surfboard before attacking the boy. Adam fell off the board under the impact of the shark's blow. He flailed about in the waves, defenseless, as the shark turned to attack again.

Despite Adam's attempts to hit the shark on the snout, the shark's teeth found their mark. The boy was bitten twice, once in the lower abdomen and once on the right hip. Adam managed to call out to his friends, "I've been bitten by a shark. I'm bleeding." As he lay in the water, he doubted that he could defend himself from the shark's next attack, for he was growing weak as a result of shock and loss of blood.

But then the dolphins came to the rescue. Jason Moloney recalled the scene later: "When Adam looked down there was blood everywhere. He freaked out and started bashing madly at the water and hitting out at the shark to stop it coming back. While Adam was bashing madly at the water, the dolphins also started splashing around trying to frighten the shark away."

Aggressive dolphins

Two of the dolphins broke away from the rest of the school and attacked the shark, chasing it away by butting it with their beaks. Bradley Thompson recalled what happened next: "...the dolphins move[d] in towards the shark's fin and there was a lot of splashing around. The fin disappeared and it was obvious that the dolphins had scared it away." The injured teenager managed to clamber back onto his surfboard and paddle in to the beach. From there a helicopter air-lifted him to the nearest hospital, where an emergency operation was performed. Adam's father, Terry Maguire, credits the dolphins with saving his son's life: "Forget the dog as man's best friend. Give me the dolphin any day."

Adam adds, "I'll always have a scar. But it will serve to remind me of what those dolphins did. I reckon I owe them."

Account reported by Reuters International News Agency and published in the Daily Telegraph (London), January 4, 1989, and in the Daily Mail (London), January 5, 1989.

> "When Adam looked down there was blood everywhere. He freaked out and started bashing madly at the water and hitting out at the shark."

LOST AT SEA

Zachious Benga from Sierra Leone, on the west coast of Africa, was out fishing for tarpon with the other men from his village one day when tragedy struck.

Because its wood was still green, his one-man dugout canoe, newly made from a single tree trunk, began, imperceptibly, to dry out in the merciless heat of the sun. In the late afternoon, the fishermen began to leave the fishing ground, with Benga one of the last to paddle in the direction of his home at Mama Beach. The other canoes were far ahead of him when he heard a crack: the canoe had split from stem to stern.

Drifting toward death

Benga pulled himself onto one half of his canoe. He called frantically to the other fishermen, but they were too far away to hear. Then he tried to paddle toward the shore, but the current was too strong and pulled him far out to sea.

Night fell and the water around the exhausted fisherman was filled with the sounds of splashing. It seemed to Benga that this was the sound of sharks' fins breaking the still water. Terrified, he pulled himself higher onto the remains of the canoe and forced himself to stay awake throughout

With dolphins nearby he knew that sharks would not dare approach. All day the dolphins stayed by his side.

the night. For he knew that if he relaxed his grip and slipped back into the water, the sharks would surely devour him.

At first light, Benga was finally able to see that the splashing was caused not by sharks but by a school of dolphins. The fear that had held him in its grip all night abated. With dolphins nearby he knew that sharks would not dare approach. All day the dolphins stayed by his side as he drifted aimlessly with the current.

Vigilant friends

Night fell again with no sign of land or a passing ship. Benga was exhausted. Almost delirious from hunger and thirst, he kept drifting off to sleep. Yet each time he began to lose consciousness, the dolphins reportedly splashed him with water to keep him awake. All through the night they kept up their vigil, waking him each time he dozed.

Zachious Benga was glad to see the morning light, but he knew that he could not survive another day without water. The sun's burning rays brought on an intolerable thirst. As he began to prepare himself for death, he realized that he was alone in the vastness of the ocean. The dolphins had disappeared. Then, just as he lost consciousness for what he feared would be the last time, he heard voices. A passing ship had spotted him.

Returned from the dead

Benga was taken to a nearby port, where he gradually recovered from his ordeal. Thirty days after being lost at sea, he returned to his native village, where his family and friends had given him up for dead. There he told the story of how he had been saved by dolphins.

Account published by Horace Dobbs in **The Magic of Dolphins** *(Lutterworth Press, England, 1990).*

AIMLESS AVIATORS

"During World War II, six American aviators shot down by the Japanese were crowded together in a rubber dinghy and had lost their oars. A dolphin appeared, and pushed the dinghy for several hours to the sandy beach of an island, from which they were rescued."
George Llano, Airmen Against the Sea *(1955)*

A CRY FOR HELP

In 1988, four French fishermen were sailing off Antibes on the French riviera. They became witnesses to a strange episode that is apparently unique in the history of human-dolphin interactions.

A strange visitor

On that autumn morning, a dolphin surfaced beside their boat. There is, of course, nothing unusual in seeing a playful dolphin frolicking beside a boat in the Mediterranean. These delightful creatures have been a familiar sight ever since humans first took to the sea in ships. But there was nothing playful about the behavior of this dolphin. Rather, the dolphin seemed to have a purpose. The creature kept nudging the boat, then swimming away from it, and returning to nudge it again. At last, the boat's captain, 66-year-old Vincent Galazzo of Nice, realized that the dolphin wanted the boat and its crew to follow him. Since they had finished their day's work, they did so.

The captain's story

Later, Galazzo recalled what happened next: "He was healthy but obviously distressed and was trying to attract our attention to something. We started the engine and followed him for several hundred meters. He knew exactly where he was taking us. Ahead of us was a badly injured female dolphin barely floating in the water. Her dorsal fin had been almost torn off and there was blood on the water. The male raced toward her. He was using his own body to keep her from sinking. He pushed her toward us, crying out all the time. She had probably been hit by a speedboat. The wound was very deep...and as we looked closer, we realized sadly, she was already dead.

Without hope

"But the male wouldn't seem to accept it and was obviously hoping we could save her. It was impossible, though — even if she had been alive, we couldn't have done a thing.

"You could hear him crying and see the blood on the water. And you knew the sharks would come for her. There was simply nothing we could do. Eventually, we started up the motor again and left. I've been going to sea for many years...and this was the strangest and saddest thing I've ever seen."

Account published in the **Mail on Sunday** *(London), October 2, 1988.*

A dolphin surfaced beside their boat. The creature kept nudging the boat, then swimming away. At last, the captain realized that the dolphin wanted the boat to follow him.

SHEPHERDS OF THE SEA

"Four spear fishermen, lost in thick fog amidst treacherous rocks, saw four dolphins nudging their boat to the left. Moments later, they found they had just missed the rocks. This happened again, so the men decided to follow the dolphins. Half an hour later, the dolphins started circling the boat, so the fishermen just waited there. When the fog finally cleared, they found they had been led safely to anchor in a bay."

South African Digest,
June 1978

A WHALE OF A JOB

"A school of dolphins has helped to save three stranded whales after human rescue attempts failed. Three humpback whales were trapped at low tide behind a sandbank off the east coast of the Gulf state of Oman. Airmen spotted the whales from a helicopter on Friday and sent rescue boats in twice to try to shepherd them out to sea. But the whales remained stuck. Then the dolphins joined in the rescue attempt, and the whales were guided out at high tide."

Daily Express (London),
December 26, 1988

COMMENT

The question of why dolphins behave so helpfully is one of many mysteries of the natural world. We know that these creatures are highly intelligent and able to communicate with each other, but such tales have led some people to believe that they have a well-developed sense of altruism. But is that true? Most marine biologists don't think so.

Instead they suggest that most people read too much into stories of rescues at sea by dolphins. It is more likely, they say, that such rescues are just part of a range of instinctive behaviors that dolphins use to help their own species survive. Thus humans in danger at sea are no more than the lucky beneficiaries of these deeply ingrained dolphin survival techniques.

Dolphins are, of course, intelligent. They can communicate with others of their species. They are sociable animals. But they are probably not trying to help humans when they ward off sharks who attack them. According to Dr. Margaret Klinowska of Cambridge University: "Dolphins do seem to have an instinctive behavior to attack sharks, ramming them with their beaks. But they do it to protect themselves rather than to protect human beings." The British naturalist Sir David Attenborough believes that dolphins may help humans in distress "because humans are approximately the same size and they may feel they are helping others of their own kind."

Coastal pilots

It is harder to explain why dolphins sometimes push boats into safe harbors or keep shipwrecked sailors company, as they did in the case of Zachious Benga. Perhaps dolphins who shepherd humans are merely doing what comes naturally to sociable animals who spend most of their time in large schools.

But humans are not the only creatures dolphins help. There have been a number of reports of dolphins helping whales. In two separate incidents in New Zealand, large herds of beached pilot whales have been refloated at high tide by teams of human volunteers. Whales often beach themselves again after being refloated, but in both these cases, schools of dolphins appeared on the scene and guided the whales to safety.

Navigating along magnetic lines

Some experts say that whales navigate by sensing variations in the earth's magnetic field as a series of contours. Sometimes these contours run into the shore instead of parallel to it. Many strandings of live whales occur at these points along the coast. And strandings usually involve whales who prefer deep waters. Perhaps this is how dolphins, who often live along the coast, can help. But why they choose to do so, no one knows.

The case of the dolphin who sought assistance from the fishermen is by far the oddest of the stories recounted here. Dolphins do talk with one another in

---

In both cases, schools of dolphins appeared on the scene and guided the whales to safety.

---

an underwater language of clicks and whistles. But they have rarely been seen communicating so clearly with humans.

The emotions that dolphin seemed to be experiencing could almost be said to be human. The display of love and grief that moved the hearts of the fishermen is the hardest part of the story to explain. And yet dolphins do exhibit other forms of "human" behavior: play and a sense of humor. Dr. Peter Evans of Oxford University, however, questions whether we are right to assume that we are the most advanced life-form, and whether dolphins are comparable to humans at all: "If you put an animal on a pedestal — and they surely deserve it — why does it have to be next to us?"

CHAPTER ONE

SECRET ANIMALS

The natural world contains many fascinating and curious creatures. Some show remarkable talents; others possess amazing physical characteristics. And there may well be more strange animals that have yet to be discovered.

There is much about the animal world that remains hidden. Is it possible that animals long thought to be extinct may still exist? And could it be that some creatures, having lived for thousands of years, have escaped scientific detection right up to the present day? And are there forms of animal behavior that humans have never seen before?

These are but a few of the unanswered questions that haunt any comprehensive examination of the animal world as it exists today. Clearly, certain animals do

exhibit unexpected and even bizarre talents. For the natural world apparently contains animals with the power to hypnotize, and animals that can make sounds not usually associated with their own species, such as mice that can sing.

Many intriguing questions abound. For how does science explain the existence of such unnatural animals as green polar bears, striped cheetahs, and spotted zebras? Do such phenomena as rat kings and glowing owls really exist? And is it possible that the beasts described in seemingly fanciful eyewitness accounts are in fact the modern descendants of creatures that were thought to have died out thousands of years ago? It is certain that many mysteries of the animal world remain to be deciphered and explained.

Unusual okapi
An early 20th-century painting clearly shows the striking patterns of the okapi's coat. The discovery of this ornate creature in 1901 caused a scientific sensation; since its discovery it has been successfully bred in captivity.

New species
Some people believe that the 20th century has revealed few previously undiscovered animals, or few that have been rediscovered after long periods of presumed extinction, and so no new species remain to be found. In fact, this is untrue, for a varied array of spectacular newly discovered and rediscovered animals has been documented since the early 1900's.

One of the most famous is the okapi, which first attracted Western attention when it was referred to in journalist Henry Morton Stanley's book *In Darkest Africa* (1890). Stanley talked of a "donkey" hunted by the Wambutti

pygmies in the Ituri Forest in central Africa. The first tangible evidence for the existence of this beast, known to the Wambutti as the okapi, came ten years later, when the governor of the Uganda Protectorate, Sir Harry Johnston, visited the Belgian fort at Mbeni in the Congo Free State (now part of Zaire) and was given two striped headbands made from an okapi skin. The stripes did not match those of any known animal. Johnston

---

An array of spectacular newly discovered and rediscovered animals has been documented since the early 1900's.

---

asked the Mbeni officers to save him the next okapi skin that they obtained.

In 1901 Johnston received a skin and two skulls. He submitted these to the eminent British zoologist Prof. Edwin Ray Lankester at London's Natural History Museum. After examining the remains, Lankester announced that the okapi was a short-necked, forest-dwelling giraffe, with a reddish coat and black and white stripes on its rump and legs.

The discovery of the okapi was a scientific sensation but it was far from being the last major surprise from Africa. In 1898 the British explorer Ewart Grogan came across the skeleton of a gigantic ape, larger than any ape he had ever seen, in the region of the Virunga volcanoes in East

Do not disturb
Mountain gorillas sometimes put on ferocious displays when disturbed, but studies show that they are normally shy, unaggressive creatures.

Africa along the borders of present-day Rwanda, Zaire, and Uganda. This large skeleton appears to be the first record of the mountain gorilla. If Grogan had sent it to England for examination, he would have been credited with the animal's discovery. But the official recognition of the mountain gorilla was made four years later, in 1902, from a skin specimen collected by Oscar von Beringe, a captain in the Belgian Army. The subspecies, which is the largest known of the apes, is now named after him (*Gorilla gorilla beringei*).

Ape hug

Until this time, only one, smaller, subspecies of gorilla was known. Its natural habitat stretched from Cameroon to the lowlands of Zaire. Yet the local tribal peoples had been telling visitors tales of an enormous mountain-dwelling ape that reputedly hugged women so hard that it could kill them. Such stories were dismissed by Western scientists as absurd legends until Beringe shot his specimen in 1902.

Experts now describe the mountain gorilla as rare — there are only between 5,000 and 15,000 of these animals left. Reasons for the apparent decline in their population include destruction of their habitat and demand for specimens by medical institutions and zoos.

The Congo peacock

One of this century's most important newly discovered birds also comes from Africa. In 1913, Dr. James Chapin, an ornithologist at the American Museum of Natural History, was participating in a scientific expedition to the Ituri Forest in the Congo (now Zaire) when he saw a native headdress containing a feather that he could not identify. Chapin took the feather home and showed it to other ornithologists. The experts spent a considerable time attempting to learn what mysterious species the feather had come from but to no avail. Finally, Chapin placed the feather in his desk,

A very plain peacock
The Congo peacock is the only true member of the pheasant family in Africa. It is very primitive, lacking the beautiful fan-like train of the more familiar Asian peacocks.

A new type of hippo
Until 1913, when it was discovered to be a separate species, the pygmy hippopotamus had been believed to be either extinct, or a juvenile form of the common hippo. The pygmy hippo is only one-tenth the weight of the common hippo, measuring 2 feet 6 inches high and 5 feet 10 inches long.

Bilkis gazelle
In 1985, mammalogists Dr. Colin Groves and Dr. Douglas Lay brought a new species to the attention of science. This was the Bilkis gazelle from North Yemen, named after Bilkis, the Queen of Sheba, who was once believed to have ruled over much of what is now North Yemen.

The world's largest pig
The giant forest hog, whose total length sometimes exceeds seven feet, is the world's largest species of wild pig. The existence of this grotesque creature was confirmed from a skull and an incomplete skin found in Kenya in 1904.

Protected species
The takahe is the official bird of the Ornithological Society of New Zealand. Rediscovered in 1948, this rare bird is now under the protection of the New Zealand government.

and there it remained, still unidentified, until 1936. During that year, Chapin visited the Congo Museum in Belgium, and on top of a cabinet he spotted a pair of stuffed birds that had been sent to the museum in 1914. The wing feathers of the female bird were identical to those of the specimen that had puzzled him for 23 years. A Belgian friend later told him about a strange bird the friend had eaten in the Ituri Forest; the sketch he drew of the bird looked exactly like the stuffed specimens.

Out of Africa

It became clear to Chapin that a type of pheasant resembling a peacock lived in the Ituri Forest. This, the first and only member of the peacock sub-family from Africa, Chapin named *Afropavo congensis* (which means "Congo peacock") in 1936. Since then the Congo peacock has been found in many parts of the Congo forests and has even been imported into several aviaries outside Africa. It has been bred successfully in captivity since 1960.

Another unexpected ornithological event was the surprising rediscovery of a rainbow-feathered, flightless rail from New Zealand. In 1847 Walter Mantell, a naturalist from New Zealand, discovered several bones in the country's North Island that seemed to belong to a very fat rail. He sent the unusual

bones to the British Museum, where they were examined and classified as bones of *Notornis mantelli*, or the takahe.

Hide-and-seek

Soon after, more bones reached the British Museum from the South Island; these too belonged to the takahe. After these discoveries, the takahe appeared to play hide-and-seek over the years, with specimens and tracks popping up at

In 1894 the first live takahe ever recorded in modern times was caught on the North Island.

different times in various parts of the islands. In 1894 the first live takahe ever recorded in modern times was caught on the North Island. Although its skin was kept for many years, it has now been lost and the only evidence of its existence is in written records. Fortunately, another

Birds of beauty
A 19th-century engraving of American ivory-billed woodpeckers. These birds were sighted in the 1960's after years of presumed extinction.

takahe was captured alive in 1898 on the shore of Lake Te Anau on the South Island. On this occasion, the remains were preserved and are kept in the Dunedin Museum to this day.

Population growth

Then half a century went by without a single authenticated sign of the enigmatic bird, and it was generally agreed to be extinct. But in November 1948 a search led by British physician Dr. Geoffrey Orbell found two living takahes around the shores of Lake Te Anau. Before the year was out, Orbell had persuaded the New Zealand government to declare a prohibited zone in the South Island to protect the areas in which the birds were found. Today, it is estimated that the full population numbers between 200 and 300 birds.

According to traditional zoology, the only big cats in Mexico are the puma and the jaguar. But for more than 300 years the people of Sinaloa on Mexico's western coast have claimed that there is a third type of cat, the onza. They claim that the onza looks like a puma but with a more slender body and longer limbs. Scientists discounted such reports, however, as misidentifications of either pumas or jaguars.

Yet by the 18th century, the onza was believed by many people in Mexico to be the country's third type of big cat. In 1757 a Jesuit missionary, Father Ignaz Pfefferkorn, posted to the little-explored north Mexican province of Sonora, referred to an animal that resembled the onza, describing it as being far less timid than the puma but of a similar shape except for its longer, thinner body.

And a similar animal was described during the 19th century. But since there was a distinct lack of physical evidence to examine, zoologists ignored such accounts, dismissing the onzas as nothing more than misidentified pumas.

More sightings were reported in the 20th century, and following the discovery of a skull believed to belong to the onza, the mystery cat was brought to the attention of J. Richard Greenwell, Secretary of the International Society of Cryptozoology. Greenwell began his own investigations and in 1985 he located two more onza skulls in the province of Sinaloa.

The onza's tale
This 1961 book by long-standing onza enthusiast Robert E. Marshall underlines the public interest in the story of Mexico's enigmatic big cat.

Clues from the skulls

All three skulls were examined in detail by the German big-cat researcher Dr. Helmut Hemmer from the University of Mainz. This was because Hemmer had suggested that the onza might be a relative of a type of cheetah that inhabited parts of North America about 10,000 years ago. But to determine whether or not Hemmer's theory was correct, the scientists needed a complete specimen of the animal for examination. Yet they realized that to locate a creature as elusive as the onza could take as long as a lifetime. In fact, thanks to a stroke of luck, it took less than six months.

Searching for an identity

On January 1, 1986, an onza was shot in Sinaloa by rancher Andres Rodriguez Murillo, who feared that it was about to attack him. A scientific team headed by Greenwell undertook a careful examination of its carcass, and the zoological world is awaiting the results.

The search is now on to discover the onza's true identity. Several possibilities have been suggested. These include the existence of a mutant puma, a specialized puma subspecies, or an altogether new species. However, only more time and research will tell.

THE FOREST GOAT

Perhaps the latest major zoological success story is currently taking place in Vietnam. In 1992 a research team headed by scientist Dr. John MacKinnon searched the country to locate areas that were in need of protection from exploitation and destruction. During the expedition, an amazing zoological find was unearthed in the Vu Quang nature reserve, not far from the town of Vinh. The team found three sets of long dagger-like horns, one with part of the skull attached, which seemed to belong to an unknown antelope-like creature.

Camera traps

The local hunters call this beast — a specimen of which has still not been captured — the forest goat. MacKinnon says that the creature, which may soon be declared a newly discovered species, is most likely to be a relative of the Sulawesi anoa, a tiny water buffalo that resembles a deer. But to work out exactly what sort of animal it is, MacKinnon needs to examine a specimen or see a photograph of the creature. At present, plans are underway to set up camera traps to catch the animal on film.

Forest finds
John MacKinnon exhibits strange horns believed to belong to a newly discovered species.

23

ANIMAL HYPNOTISTS

Eyewitnesses claim that some animals can mesmerize, or "fascinate," other animals as part of their hunting strategy.

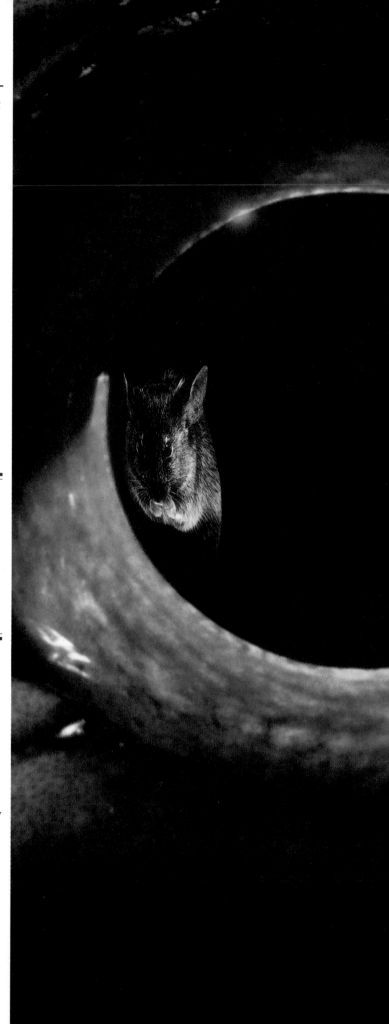

NAKES ARE THE ANIMALS most commonly credited with hypnotic abilities. In his book *Three Visits to Madagascar* (1858), English missionary William Ellis describes watching a snake hypnotize a mouse. "The snake fixed its eye on the mouse, which then crept slowly towards the snake, and, as it approached nearer, trembled and shrieked most piteously, but still kept approaching until quite close, when it seemed to become prostrate, and the snake then devoured it."

Hypnotic explanations

Although Ellis believed that the mouse had indeed been hypnotized, such behavior may have a simpler explanation. Snakes often bite their prey and wait for it to die before devouring it, all the while watching out for any other animal that might attempt to steal the dying creature. Some skeptics therefore claim that the

> "The snake fixed its eye on the mouse, which then crept slowly towards the snake...until quite close, when it seemed to become prostrate, and the snake then devoured it."

victim's odd behavior is due to the effects of the venom, while the snake's fixed gaze is simply due to the fact that snakes have no eyelids and cannot blink.

In a 19th-century report in the journal *Zoologist*, the English clergyman Henry Bond recalled noticing a hedge sparrow in a bush emitting shrill cries of alarm and hopping down from twig to twig toward a coiled snake that was gazing fixedly at it from the base of the bush. The bird seemed irresistibly drawn toward the snake. Bond's approach frightened the snake, however, which then slithered away. In this case it seems unlikely that the prey had been injected with venom, because it was reported to have immediately flown away.

Mammalian mesmerizers

Stoats and weasels are also reported to be able to hypnotize their prey. Once again Henry Bond was a witness. Walking in Somerset, England, he suddenly heard a cry of distress and spotted a rabbit running around in an ever-decreasing circle. The rabbit moved with a curious halting gait. Coming closer, Bond saw a

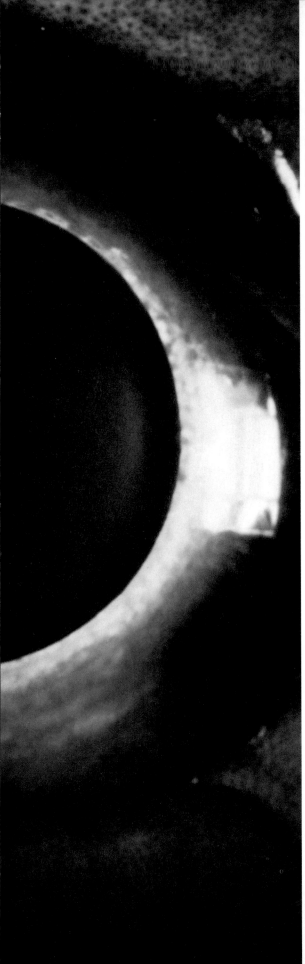

stoat at the center of the rabbit's circuit, continually turning its head to keep its gaze fixed on the rabbit. Just as the rabbit was nearly within reach of the stoat's jaws, Bond hurled a stone at the stoat. Although he missed, the stoat's spell was broken and the rabbit was able to escape to safety.

Many theories have been put forward to explain the hypnotic power of certain animals. But it seems likely that these are not cases of hypnotism at all. Rather, the odd behavior of the "hypnotized" animal is probably the result of sheer terror.

The dance of death

Some animals appear to stupefy their prey by performing a series of strange movements before them. Weasels, for example, have been seen using a technique known as the dance of death to catch birds. On February 3, 1919, in Hampshire, England, nature lover Albert Rowland saw a weasel apparently dancing in the snow while a rook sat and watched it. Running rapidly around in small circles, leaping and twisting as it did so, the weasel gradually edged nearer until suddenly it sprang at the bird. Although the weasel missed, it repeated the procedure several times. The rook made no attempt to fly away and finally the weasel caught it.

Solo dancing

Some researchers suggest that this dance of death is a deliberate hunting technique, the predator's movements designed to intrigue the prey and render it defenseless. However, weasels have been seen dancing even when no other animals are present. This has led to the belief that these movements may be an involuntary response to intense irritation, possibly caused by parasitic worms inside the weasel's skull.

Whatever the reason for these dances, it seems probable that rather than succumbing to any hypnotic influence, the weasel's prey may simply become so confused that it fails to notice its enemy closing in for the kill.

ANIMAL ACUPUNCTURE?

In an article in the journal *Tigerpaper* in 1986, Indian naturalists Raza Tehsin and J. S. Nathawat described a method used by livestock thieves in India's Kumbhalgarh region to immobilize sheep or goats. The thieves pulled the animals to the ground and put small weights on their ears.

Testing the technique

The two naturalists tested the technique by using lightweight stones to immobilize an antelope, two rodents, and a young deer. Since tigers and other big cats normally bite with one of their canine teeth near the back of the ear of their prey when seizing it to pull it down, Tehsin and Nathawat speculated that the animals might be practising a form of acupuncture. The predatory cats may deliberately bite this specific area of their victim's body in order to induce paralysis, thereby allowing them to bring the animal down more easily. This idea is still only a theory, and has yet to be confirmed.

Lioness attacking wildebeest

SERPENT RIDDLES

Can snakes hear without ears, leap without legs, and speak without vocal cords? And can they reach the terror-inducing proportions reported by eyewitnesses in some of the world's remotest corners? These and other serpent riddles remain undeciphered.

ACCORDING TO THE DICTATES of serpent anatomy, snakes should not be capable of jumping — they have no legs and move forward by pushing backward against a stationary object such as a twig, stone, or lump of earth.

But people in many countries claim to have seen jumping snakes. The zoologist Dr. Maurice Burton devoted an entire chapter to eyewitness reports of jumping snakes in his book *More Animal Legends* (1959). In Burma, people believe that the king cobra

---

In Burma, people believe that the king cobra can, when angry, propel itself forward in a series of speedy, whiplike jumps.

---

can, when angry, propel itself forward in a series of speedy, whiplike jumps; while in Greece during the First World War, encounters with leaping snakes were reported by soldiers in the trenches at Thessaloníki.

Fear and aggression

Fear seems to motivate another snake to show off its jumping skill. A letter reprinted in Burton's book from M. F. Kerchelich of Sarajevo, Bosnia-Herzegovina, claims that Dalmatia, on the coast, and the nearby mountainous regions of Herzegovina and Montenegro harbor a type of jumping snake called the *postkok* (meaning "jumper"), which is poisonous and will leap into the air when frightened or in an aggressive mood.

Jumping snake

The postkok is a long thin snake that varies in color from granite to a dark reddish brown and often suns itself on rocks or roads, camouflaged by its coloration. On one occasion, Kerchelich said that he saw a snake curled up in the road asleep. Perhaps startled by the presence of a human being, the snake leapt up and jumped nearly three feet in the air, landing five feet away in a ditch. On another occasion, a postkok apparently jumped up at Kerchelich's car, hitting the front bumper before being crushed by the rear wheels.

Writing in the sand
As this sidewinder glides across the Namib Desert, Namibia, it leaves a series of lines that make it look as if it has jumped sideways across the sands.

Col. Percy Fawcett

LOST IN THE JUNGLE
In the early years of the 20th century, Col. Percy Fawcett explored the uncharted reaches of the Amazon jungle in South America. In 1907 he shot a huge anaconda that was making its way up the bank of the Rapirrãn River in Brazil. Fawcett claimed to have obtained a measurement of 45 feet for the portion of the snake that lay on land, and he estimated that 17 feet more remained in the water, thus making a total length of 62 feet.

Lost explorer
Fawcett was not able to bring back evidence of his discovery, and the river swallowed up the body of the snake. In 1925 Fawcett himself disappeared into the jungle. A skull that some say is his was found in 1951.

But despite the numerous eyewitness accounts of this phenomenon, most scientists do not believe that snakes can jump. One supposed jumping snake, the *springslang* (whose name literally means "jumping snake") of the Kalahari Desert and Namibia, proved not to be a snake at all but a species of near-legless lizard. As Burton's book shows, however, reports of jumping snakes are not limited to southwestern Africa — many sightings have come from regions that are not believed to contain any legless lizards.

Record breakers
We may be thankful that the longer the snake, the less likely it is to jump — if snakes really do jump — since sightings of gigantic serpents, far longer than any scientifically verified specimens, have often been reported in remote regions of Africa and South America. What brings even the most fantastic of these reports into the realm of the possible is the ability of snakes to continue to grow, albeit more and more slowly, throughout their lifetime. If a snake lived long enough, and some are known to have lived 40 years, it might, theoretically at least, approach some of the enormous lengths that have been reported.

In August 1959, Belgian aviator Col. Remy van Lierde encountered an ultra-large specimen while flying a helicopter at an altitude of 500 feet above a patch of vegetation and some termite hills in the Katanga region of the Belgian Congo (now Shaba, Zaire). He saw an enormous snake emerge from a hole in the ground and rear upward, directly toward his helicopter. Later, experts estimated that the snake was about 200 feet long.

American anacondas
In South America, many huge anacondas have been spotted, and some of these great snakes have even been shot and photographed. On January 24, 1948, for instance, a Brazilian newspaper, *Diario de Pernambuco*, published a photograph of

---

In 1912 a reticulated python measuring 32 feet 9³/₄ inches was shot on Sulawesi.

---

an immense anaconda that had recently been killed with a machine gun near Manaus, on the Amazon River. This monster was said to measure some 130 feet in length, and 2 ½ feet in diameter.

On April 28, 1949, another Brazilian newspaper, *A Provincia do Para*, published a photograph of what was allegedly a massive anaconda floating on the Abunã River in the territory of Guaporé; the photographer claimed that

Great lengths
Seven men struggle to hold a large anaconda in the jungles of northern Brazil, showing just how difficult it is to collect such a large specimen.

the snake was about 147 feet long. But because it is easy to tamper with photographs, they cannot be accepted as definitive proof of the existence of monstrous serpents.

Great lengths

But the longest authenticated snake on record is only a fraction of the length of these monsters. In 1912 a reticulated python measuring 32 feet 9 ³/₄ inches was shot on the Indonesian island of Sulawesi. No snake submitted for scientific examination since has even approached that length. This may not be due so much to the fact that longer snakes do not exist as to the difficulty of hauling the putrefying remains of a dead python or anaconda back to civilization.

Stinging tails

Some snakes are unique by virtue of their distinctly unsnakelike shape, not their size. One of North America's most infamous mystery snakes is the horn snake — widely reported during the 18th and 19th centuries. It is said to resemble a rattlesnake, but its tail is equipped with a sharp, horn-like projection instead of a rattle. This horn is supposedly endowed with deadly poison, and the snake is said to use it to kill prey just as a scorpion uses its tail to sting. So far, no specimen has ever been submitted for scientific examination. Some people therefore believe that rather than being a separate species, the creatures identified as horn snakes are known species that have simply been misidentified.

The red-bellied mud snake *Farancia abacura,* of Louisiana, is the most likely candidate for such misidentification because its tail bears a long and spiny (albeit nonpoisonous) scale. When threatened it will actually attempt to stab its foe with this scale. The venomous water moccasin and the copperhead have also frequently been mistakenly identified as the horn snake. It has even

been suggested that some horn snakes may have been old rattlesnakes whose rattles had become eroded to a point.

Living legends

Perhaps because of their strange habits and appearance, snakes are often the subject of legends and folktales. Yet according to cryptozoologists, some of the serpents famed in folklore may be real, but undiscovered, species.

In 1987, Swiss cryptozoologist Michel Dethier and his Japanese wife Ayako Dethier-Sakamoto brought the elusive *tzuchinoko* to widespread attention. It is an unusually short, thick-bodied, horned snake from Japan. Although it has been reported for centuries by eyewitnesses, it has never been positively identified by scientists. The presence of small facial pits, horn-like projections above its eyes, a well-delineated neck, and a triangular-shaped body when viewed head-on, suggest that the tzuchinoko may be a strange mutant of the Halys pit viper. On the other hand, it might

Open wide
After swallowing the gazelle it has killed, this python may not need to eat again for several months. The snake's anatomy allows it to swallow prey far larger than itself — the jaw bones are held together with muscles and ligaments, allowing the mouth to stretch open. The teeth point backward to make it easier for the snake to drag the prey into its mouth.

The real horn snake?
Calmly coiled, the red-bellied mud snake of Louisiana gives no hint that it can lash out with the scaly tip of its tail if threatened or disturbed by an attacker.

Super-loud hiss
When air rushes past the enlarged flap covering the North American bull snake's windpipe, its hiss is amplified so that it produces the bull-like grunting sound that gives it its name.

SNAKE STONES

In many parts of the world, snake stones are used to treat recently inflicted snake bites.

In one incident that occurred in Sri Lanka in the 19th century, a British government officer saw a Tamil being treated for a snake bite on his finger with two small, polished black stones. When the stones were applied to the wound, they stuck fast, soaking up the blood oozing from the bites.

For three to four minutes, the wounded man's companion rubbed his arm downwards. At length, the snake stones dropped off of their own accord. The witness reported that the suffering of the man then appeared to subside and he went on his way.

Despite many reports like this one, however, scientists do not believe that stones can extract the poison from snake bites.

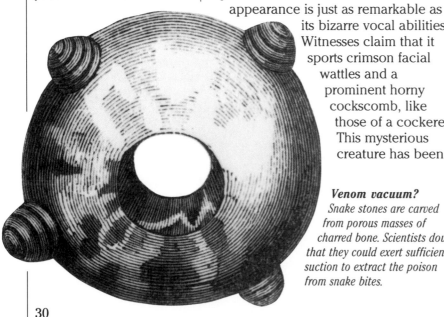

well be a specialized species in its own right, still awaiting formal recognition.

The three-foot-long tatzelworm has been reported for centuries from the remote corners of the Alps. It is a thick-bodied, stumpy-legged creature that combines a serpent's body with the legs of a lizard. If it is ever captured and classified, it will probably prove to be neither a snake nor a worm but a lizard, for its appearance is similar to a large, legless lizard known as the glass snake.

Something to crow about

Snakes do not have ears or vocal cords, which would seem to imply that they are unable to hear or make sounds. But there are many recorded instances, some well documented, of snakes that make distinctly unsnakelike sounds. The crowing crested cobra, for example, is reported to be able to crow. The snake's appearance is just as remarkable as its bizarre vocal abilities. Witnesses claim that it sports crimson facial wattles and a prominent horny cockscomb, like those of a cockerel. This mysterious creature has been

Venom vacuum?
Snake stones are carved from porous masses of charred bone. Scientists doubt that they could exert sufficient suction to extract the poison from snake bites.

seen in various locations across tropical Africa. The September 1962 edition of *African Wild Life* contained an account from a reader, John Knott, who had inadvertently driven over a large crested snake in the Kariba area of what is now Zimbabwe. He cautiously inspected the

---

The bull snake is said to emit bovine grunts that can be heard up to a distance of approximately 100 feet.

---

dying animal and later claimed that its crest was a perfectly symmetrical, distinctly formed structure, whose five prop-like rods appeared to enable the snake to make it erect.

Serpent sounds

In addition to the crowing crested cobra, many other snakes are rumored to be able to emit sounds other than their usual sibilant hisses. For example: According to Capt. G. B. Ritchie (*East Africa*, September 1929), at breeding time the African puff adder produces a clear, bell-like note, audible for about 200 yards; the 19th-century Scottish explorer and missionary David Livingstone learnt of a South African snake termed *nogo putsane* (meaning "kid snake") because it reportedly could produce a goat-like bleat; and in North America, the bull snake is said to emit bovine grunts (hence its name) that can be heard up to a distance of approximately 100 feet.

So far, science has only verified the ability of bull snakes to make sounds, and has generally discounted all the other reports of vocalizing snakes. Perhaps because snakes have no external ears, it is commonly believed that they are deaf to all airborne sounds, only picking up those sound vibrations that are conducted through the earth. If this were true, it would mean that these reptiles would not be able to hear the noises they produced. But zoologists have known for some time that snakes can hear. Airborne sounds are transmitted through thin plates of bone in the snakes' faces to their inner ears. These bones are sensitive to low-frequency sounds, so snakes can hear as efficiently as lizards, which are equipped with external ears.

Finding a voice

Similarly, because snakes' larynxes (voiceboxes) lack vocal cords, it is also believed that they must be incapable of emitting sounds, even though birds and many fishes, which also lack vocal cords,

Grass snake
In 1932 a report of a singing snake came from Mrs. Duncan Carse of Berkshire, England. She had startled a grass snake as it slept in her garden. The surprised snake allegedly uttered a very loud, bird-like cry of alarm, heard several yards away by her husband, who was indoors.

are accomplished vocalists. Birds, for instance, have a separate organ called a syrinx that they use to produce sounds.

In 1980 Philip Chapman of Bristol City Museum, England, discovered another snake that can make sounds. On an expedition to Sarawak's limestone caves in Malaysian Borneo, he heard a cat-like miaow coming from a coiled Bornean cave racer. This slender blue species has long been known to naturalists, but no one suspected that it could emit any sound other than a normal hissing. Clearly, there is still more to learn about snakes' vocal and aural capabilities.

ANIMAL VOCALISTS

Along with snakes, the animal kingdom contains other creatures that are known to produce surprising sounds. Mice can not only squeak but are also reported to be able to sing, while dolphins and elephants can use sound to help them survive.

Reports of singing mice abound. In his *Animal Curiosities* (1922), English naturalist W. S. Berridge referred to one specimen, kept as a pet, that could run up an octave when singing, inflating its throat like a bird and concluding with a trill. On December 15, 1936, a singing mouse called Minnie performed on a Detroit radio station. Listeners likened her voice to that of a robin or a tone-deaf canary! And in May 1937, a transatlantic radio talent contest for singing mice was staged, featuring rodent songsters from places as far apart as London, Illinois, and Toronto.

No one knows why singing mice evolved. Some scientists have suggested, in all seriousness, that they are merely normal squeaking mice with inflamed respiratory organs; others, with equal gravity, have purported the theory that they are squeaking mice with very deep voices.

Singing mouse

During the late 1980's, researchers at Santa Cruz's Long Marine Laboratory recorded loud ultrasonic sounds produced by dolphins. These sounds may solve the mystery of how dolphins, who have blunt beaks and sometimes even lack functional teeth, successfully capture their prey. Dolphins may stun (and perhaps even kill) their prey by producing ultrasonic "bangs." This noise would temporarily deafen and disorientate the prey, thus enabling the dolphins to catch it easily.

Masters of the sound waves

In 1984, wildlife researcher Katie Payne felt a low throb in the air while in the elephant enclosure at a zoo in Portland, Oregon. The rumbles were found to be infrasonic sounds produced by the elephant's larynx and emanating from its forehead. Elephants seem to use these sounds to communicate. Even when scattered over a vast area, they are able to come to one another's aid quickly if danger threatens.

UNNATURAL ANIMALS

In rare instances, the natural world produces an animal that does not appear to conform to the rules of nature. Although science can explain how some of these strange creatures might come about, the majority remain a mystery.

IT WAS A CLEAR NIGHT IN ANATOLIA, western Turkey, in 1918. John Welman saw something strange in the sky and stopped dead in his tracks to take a look at it. The curious object, which glowed brightly, swung around a clump of trees 600 feet away from where he stood and moved toward him, swiftly and silently. As it drew near, Welman was fascinated to see that the luminous form was, in fact, a bird. Thirty years later, writing in *Blackwood's Magazine*, John Welman described the curious vision: "Every feather of its plumage glittered with tiny points of light, a kind of frosted fire which...was bright enough to illuminate the branches of a tree through which it passed."

But the bird that Welman saw was by no means unique. Such creatures have a history, one that is well documented but hardly proven, dating back at least as

> "Every feather of its plumage glittered with tiny points of light... bright enough to illuminate the branches of a tree."

far as Pliny the Elder (*c.* A.D. 23–79). This Roman writer mentioned them in his treatise on natural history.

Certain animals, including fireflies, glowworms, and various deep-sea fishes, actively emit light by producing light-releasing compounds within their cells. Far more extraordinary, however, are reports of birds (and even certain mammals) that also glow.

Glowing owls

Owls are the most commonly reported glowing birds. One report, by British naturalist Sir Digby Piggott, claims that one of two glowing owls seen together in February 1907 near Twyford in Norfolk, England, was shot by a gamekeeper, who later identified it as a common barn owl. Two other eyewitnesses reported seeing a glowing owl at the same site in October 1907 and twice more in December. They claimed that the eerie glow emanated from the bird's breast, because the light was not as bright when the owl flew away from them.

During that same period, the French journal *Revue Français d'Ornithologie* published several reports of glowing birds, and by the mid-1940's Dr. W. L. McAtee

EFFECTS OF GENES

When a black cat and a ginger cat have kittens, some will be black, some will be ginger, and some will be tortoiseshell. This mixing of the two colors occurs because coat color is determined by genes.

Genetic instructions

Genes are the instructions for specific characteristics in the body. Each characteristic is determined by at least two forms of the same gene, which may be identical or different. If they are different, one form will be dominant and its effects will be seen in the animal, and one will be recessive and its effects will be hidden. If they are the same, even if both forms of the gene are recessive, the animal will show that characteristic.

Because there are many thousands of genes determining different characteristics in the

Albino peacock
This peacock has a genetic mutation that inhibits its natural colors.

body, there can be enormous variation within the same species.

Sometimes a mistake occurs, producing a mutant gene that is then passed on when the cell divides. Carrying a mutant gene can have various effects, but most are detrimental. Some genetic mutations affect the appearance of the animal, producing anomalies such as albinos.

of the U.S. Fish and Wildlife Service had accumulated many reports of the phenomenon, from owls to night herons. But where did the glow come from? Several theories have been put forward, but none is conclusive.

Phosphorescence seems to be related to the bird's feathers — witnesses have not seen it after a bird has molted. Some experts theorize that a bird might eat phosphorescent microorganisms and spread them over its body during grooming. Another theory is that such microorganisms could flourish in a bird's feathers if they got damp or dirty. But if that is true, why is the phenomenon so rare?

Sticky situation

Many experts now agree that the most likely explanation is that phosphorescent microorganisms (commonly the honey fungus *Armillaria mellea*), stick to a bird's feathers as it enters or leaves a tree hole, thereby making the feathers glow. But this does not explain why it is most commonly the breast that appears to glow and not the wings or head (which are most likely to make contact with the hole's edges). Also, some very large glowing birds, such as America's four-foot-tall great blue heron, have been seen, and these birds are hardly likely to inhabit tree holes. Skeptics suggest that the birds do not glow at all and the effect is merely an illusion caused by tricks of the light.

Mistakes and mutations

At San Diego Zoo in the summer of 1978, visitors were shocked to see green polar bears. Tests showed that algae had grown on the follicles of the bears' fur.

In this case, as with the glowing birds, external influences changed the bears' appearance. But most of nature's strange creatures appear to be the result of genetic mistakes. When nature goes wrong in this way, the results are often very dramatic — and difficult to explain.

Royal relatives
King cheetahs and spotted cheetahs live in perfect harmony in their natural habitat. Here the differences between the two types can be seen.

The South African king cheetah does not sport the cheetah's typical spotted coat but is instead flamboyantly patterned with ornate swirls, blotches, and a series of stripes running along its back. This big cat was once mistakenly thought to be a different species altogether, but breeding between these extraordinary creatures and ordinary cheetahs at the de Wildt

> When nature goes wrong in this way, the results are often very dramatic — and difficult to explain.

Cheetah Breeding and Research Centre in South Africa, during the 1980's, showed that the resplendent pattern of the king cheetah's coat is created by a mutant gene, similar to the one that causes tabby markings in domestic cats.

But as king cheetah researchers Paul and Lena Bottriell point out, the king also differs from other cheetahs in having a mane (hence its name) and bold rings around its tail. Also, all big cats possess stiff whiskerlike hairs called guard hairs that differ microscopically between species. The king's guard hairs are more similar to a leopard's than a cheetah's.

Although some genes can affect several aspects of a creature's physical makeup, the Bottriells are not convinced that a single mutant gene could produce such major differences. They suggest that perhaps the king cheetah is adapting to its environment and, over a long period of time, may evolve into a species quite distinct from other cheetahs.

In January 1968 a zebra that seemed to be a contradiction in terms was observed in the Rukwa Valley in what is now northern Zambia. Instead of being striped, it was spotted. Overall, it looked like a black colt with lines of rectangular white spots. The mechanism responsible is still unknown, but the animal has helped to settle a perennial dispute — is a zebra white with black stripes or black with white stripes? Zambia's spotted zebra suggests the latter. It implies that the stripes of normal zebras are formed as a result of inhibition of black pigment, which, in this case, had malfunctioned.

All in the genes?

Sometimes mutant genes cause coat coloration to malfunction, producing such exotic-looking creatures as the blue tiger that Methodist missionary Harry Caldwell claimed to have seen in China's Fukien province in September 1910 and the bright orange raccoons found in the Blue Hills area of Boston, which were captured and transferred to the Franklin Park Zoo during the 1960's.

Some genetic malformations are much more extreme and much more difficult to explain. Insects with extra

legs or wings have been caught; some even possess an extra leg instead of an antenna. Perhaps most bizarre of all, however, is a rare event associated with butterfly metamorphosis. Normally, the caterpillar's entire form breaks down during pupation and a totally different creature, the butterfly, develops. On rare

Little Snowflake
A mutant gene is responsible for the strange physical appearance of Little Snowflake, Barcelona Zoo's world-famous, blue-eyed, white-furred gorilla.

occasions, however, the head does not break down, and the butterfly emerges with a monstrous caterpillar head.

Throwbacks

In some cases, combinations of certain genes result in a throwback, a creature with features more typical of its distant ancestors than its parents. Millions of years ago, whales had two pairs of legs, yet their modern descendants have no hind limbs. A humpback whale with two rudimentary hind limbs, however, was caught in July 1919 near British Columbia's Vancouver Island.

Even more grotesque are reports of animal cyclopses, creatures with a single large eye in the center of the forehead. A lamb cyclops (whose head was later studied by anatomist Dr. A. M. Reese) was born in April 1935 on a farm in Pocahontas, West Virginia; and a pig cyclops, with a huge snout reportedly above its single eye, was born in England, in July 1871. The causes of such genetic malformations are not fully known.

Snake cyclops
Because of a genetic mutation, this snake has a single central eye, like the Cyclops of mythology.

MIRACULOUS CREATION

In the early 19th century, a strange experiment was performed. Some people say that it shows that the spontaneous creation of life, the subject of Mary Shelley's horror story *Frankenstein*, is possible.

In 1837 scientist Andrew Crosse attempted to create silica crystals by passing an acidic solution through a lump of iron oxide, causing a reaction that produces electricity. After two weeks, Crosse noticed tiny white protrusions on the surface of the electrified oxide. According to his reports, over the next few weeks Crosse watched the protrusions increase in number, grow legs, and eventually detach from the iron oxide and move around.

Mystery mites

Crosse classified the strange creatures as mites and tried to come up with a logical reason for their sudden appearance. At first he thought they might have arisen from eggs deposited by mites in the air, but he could not find any remains of shells. Then he considered that they might have originated in the water of the solution, but tests showed no such contamination. Finally, unable to explain their presence, he proposed that they might have been spontaneously generated from nonliving matter.

Although several other scientists reported similar results, including the eminent physicist Michael Faraday, Crosse's work was widely ridiculed. Even today, the mystery of Crosse's mites remains unexplained. Was Crosse simply a careless worker who had not achieved the necessary standards of sterilization, or could he have somehow discovered the secret of creating life? We still do not know.

RAT KINGS

Preserved in jars in a number of European museums are groups of black rats inextricably linked to one another by their tails. But no one knows how or why rats form these clusters.

RAT KINGS ARE GROUPS OF RATS whose long tails have become thoroughly entangled, bringing as many as two dozen rats together in a knot that means death from starvation for the entire group. Although the members of the rat king struggle to free themselves, the ties that bind them are too strong and the creatures soon perish. How the rats come to find themselves in this predicament remains a mystery.

What's in a name?

The very name *rat king* is also mysterious. It may be a direct translation from the French *roi de rats*, or the German *rattenkönig*, both of which literally mean "king of rats." Alternatively

> **The rats in each king were separated. Their tails were found to have been so tightly entwined that the tail of each freed rat clearly bore the impression of the tails of the others.**

the phrase could be a corruption of *rouet de rats*, meaning "wheel of rats." *Rouet* is the French word for spinning wheel — when the rat king is laid out on the ground, the rats' tails resemble the spokes of the wheel.

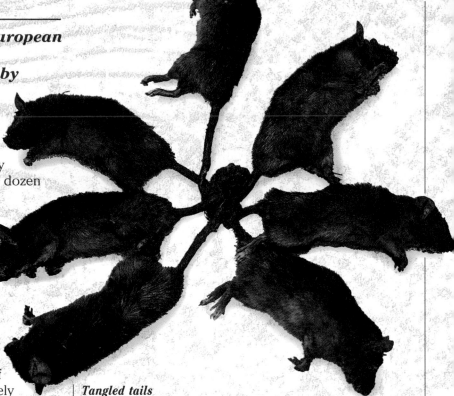

Tangled tails
Photograph of the rat king found at the farm of P. van Nijnatten at Rucphen, the Netherlands, in 1963. It shows the rat king before the knot was partially untangled and the X-rays were taken.

The largest rat king ever found, an assemblage of 28 creatures, was discovered in Döllstädt, Germany, in December 1822. Most rat kings, however, are much smaller, with only 6 to 12 members. This mighty king was one of two found in a hollow beam in a barn roof — an environment well suited to black rats, which are descended from tree dwellers and still prefer high places such as roofs and attics. All the rats in the Döllstädt kings seemed to be of the same age and size, as is typical in most rat kings. With great difficulty, the rats in each king were separated. Their tails were found to have been so tightly entwined that the tail of each freed rat clearly bore the impression of the tails of the others.

Modern rat kings

Ten well-documented rat kings have been found in this century. In February 1963, a king made up of seven rats was found by a farmer in his barn at Rucphen, in North Brabant, the Netherlands. The king was discovered when the farmer heard loud

A puzzling knot
The irregular appearance of this knot from the Rucphen rat king suggests that it was not made by tying together the tails of dead rats. The knots of man-made rat kings are generally much more regular.

squealing coming from beneath a pile of bean poles. He saw one rat and quickly killed it. But when he tried to pick it up to dispose of it, he found that it was attached to six other rats by the tail. The farmer killed the rest of the struggling rats. This king has been X-rayed and photographed in an effort to discover how rat kings are formed.

Just a hoax?

Most rat kings appear to be genuine. A few false kings were probably made in the past as fairground exhibits. Rat kings were often displayed as curiosities to the public after they were found, and a fee

Fractured tails
X-rays revealed that some of the tails in the Rucphen rat king were fractured and showed signs of callus formation, indicating that the knot had been formed some time before the rats were found.

was charged for viewing them. While tying the tails of dead rats is a simple, if unpleasant, task, it is no easy matter to knot the thrashing tails of several squirming, biting rats together. Rat kings whose members were alive when they were discovered may therefore be considered to be genuine. Yet it is not clear exactly what this means.

For despite many well-documented examples, no one knows how the rats' tails can become so hopelessly knotted. It seems most likely that the knots occur when the rats huddle together to keep warm in damp surroundings, with their tails pressed against one another. The tails may then become sticky and adhere to one another, becoming ever more entangled as the rats try to pull free.

In one experiment, one that because of its cruelty should probably not have been conducted, the tails of a number of rats were glued together. When the glue dried, the animals were released. As they struggled to get free, they climbed over one another, hopelessly tangling their tails, thereby forming an irregular knot like the ones seen in rat kings that had developed naturally.

Other animal kings

Although kings are known to occur in other species, they seem to be rarer than rat kings. Only one mouse king — a group of several intertwined young field mice discovered in April 1929 in Hölstein, Switzerland — has ever been confirmed.

A few squirrel kings have also been recorded. A seven-squirrel king was discovered in a South Carolina zoo on December 31, 1951. When the bushy tails could not be unknotted, the squirrels were separated by cutting off their tails above the knot. Two other squirrel kings have also been reported in the same zoo. Both occurred during unusually cold weather, which may suggest that the squirrels had huddled together in an attempt to stay warm and become entangled.

On September 29, 1989, a furry king made up of six squirrels was discovered by a schoolgirl named Crystal Cresseveur in a hedge outside her home in Easton, Pennsylvania. Two years later, on September 18, 1991, a five-squirrel king dropped out of a tree near an elementary school in Baltimore, Maryland.

Rattus rattus
Although not as common as its cousin the brown rat, the black rat, Rattus rattus, is the rat involved in all known rat kings, perhaps because it has a particularly long tail.

Conrad Gesner (1516–65)
The first recorded mention of rat kings was by the Swiss physician and naturalist Conrad Gesner. However, the account he gave of rat kings in his Historia animalium (1555) referred to a different phenomenon: that of a giant "king" rat attended by "a court" of smaller ones.

SURVIVORS FROM THE PAST?

Eyewitness descriptions of unidentified beasts often resemble animals that have long been extinct. But could these creatures really be survivors from the Stone Age or are they just imaginary?

*I*N NOVEMBER 1975, a strange creature was reportedly captured alive in the jungles of Kalimantan (Indonesian Borneo). It was called the tigelboat, and it was alleged to have had a tiger-like body, lion-like neck, cow-like ears, chicken-like claws, and elephant-like trunk. Since scientists believed the tigelboat to be a creature of fantasy, not fact, they never went to the prison in Tenggarong where the beast was being held to study it.

But after reading descriptions of the creature, British zoologist Dr. Karl P. N. Shuker suggested a possible identification for it in the spring–summer 1992 issue of *Strange Magazine*. Shuker believed that this beast may

---

It was called the tigelboat and it was alleged to have had a tiger-like body, lionlike neck, cow-like ears, chicken-like claws, and elephantlike trunk.

---

have been an immature tapir. Young tapirs look very little like mature specimens but a great deal like the description given of the elusive tigelboat, for they are hairy (like a lion) and striped (like a tiger), with a trunk (like an elephant), and pointed hooves (like claws).

The only tapir known on Borneo, the white-saddled Malayan tapir, died out 10,000 years ago, but sightings of tapir-like creatures are periodically reported from this vast, forested island. As noted by Eric Mjöberg in *Forest Life and Adventures in the Malay Archipelago* (1930), "It is not certain that the

The tigelboat?
This baby Malayan tapir, with its long snout, furry, striped body, and claw-like hooves, matches the description of the tigelboat that was captured in Borneo in 1975.

Primeval scene
Two palorchestids, possible ancestors of the Papua New Guinea devil pig, are shown grazing in the center of this artist's impression, along with other primitive Australian animals, all now extinct.

tapir has been met with in Borneo, although there are persistent reports that an animal of its size and appearance exists in the interior." Unluckily the tigelboat vanished from its prison, so scientists could not examine it.

Papua New Guinea's devil pigs

Curiously, unidentified tapir-like beasts have also been reported in Papua New Guinea, where they are known to the Papuans as devil pigs. The Papuans say they are dark brown or virtually black, with patterned markings, long noses, and cloven feet. These devil pigs, they claim, are about 4½ feet long.

Since this island's large mammals are mostly marsupials, it may be that if "devil pigs" do exist they are not genuine tapirs, but marsupial equivalents. Marsupials are primitive mammals who give birth to live young and carry them in pouches until they mature; the kangaroo and the opossum are examples. In some parts of the world, such as Australia, where marsupials are common, marsupial versions of a number of different mammals, including the wolf and the tiger, as well as the tapir, once existed.

Marsupial tapirs are known as palorchestids. They belong to a family of large, plant-eating marsupials called diprotodonts, which grew as large as rhinoceroses. Palorchestids may have existed in Australia until as recently as 6,000 years ago (and been contemporary with early man), but no fossil remains have been unearthed in Papua New Guinea — so far.

Yet prehistoric stone carvings found in Papua New Guinea's Ambun Valley portray a strange-looking, long-nosed animal. The mammalogist J. I. Menzies of the University of Papua New Guinea suspects that these carvings depict the tapir's long-extinct marsupial equivalent, the palorchestid. If his theory is correct, then the occasional sightings of Papuan devil pigs might prove to be some form of surviving palorchestid.

Statue stamp
This stamp shows one of the stone carvings found in the Ambun Valley of a mysterious tapir-like beast. It was issued in 1984 to commemorate the discovery of these Stone Age carvings in Papua New Guinea.

Armor-clad creatures
The glyptodont, a huge relative of the armadillos that officially died out several millennia ago, is similar to the legendary Brazilian minhocão. Witnesses claim that the minhocão is a tunneling animal with a pig-like snout whose armor-clad body is at least 13 feet long.

BARELY BELIEVABLE

The only modern-day bear officially believed to exist in South America is the spectacled bear found in the Andes of Chile and Bolivia. But northern Peru's Gran Pajonal Forest may house a much larger, all-black species (and possibly a red-furred variant too), known locally as the milne. In his book *The Rivers Ran East* (1954), explorer Leonard Clark claimed that he shot a milne there, but it fell into a piranha-infested river. Not surprisingly, he was unable to retrieve the body.

It's A....A....Milne!

Skeptics, however, find it hard to believe that the Spanish-speaking people of the region would use such an English-sounding name, one which is also the surname of A. A. Milne, the author of the children's book *Winnie-the-Pooh* (1926), in which he created one of the most famous bears of all, Pooh. Cryptozoologists have to be constantly on their guard against plausible stories of mysterious animals that might be hoaxes or misidentifications.

A detail from the cover of Winnie-the-Pooh

Another supposedly long-extinct animal that may still survive is the giant short-faced bear. The Russian scientist Dr. Nikolai Vereshchagin believes that the elusive irkuiem — a huge beast found on the Kamchatka Peninsula in eastern Siberia — may be a surviving giant short-faced bear.

Bear facts

Local hunters say that the irkuiem resembles an enormous polar bear and estimate that it is five feet high at the shoulders and weighs 1 ½ tons. They claim that they have shot one of these rare creatures about once every five years. One hunter sent a photograph of the bear's skin and samples of its fur to zoologists in Moscow and St. Petersburg. But until a skull or some teeth have been collected, zoologists will continue to be unconvinced that the creature exists. Giant black bears apparently have also been seen in this region, but these now appear to have died out.

Canada was once home to another enormous bear called MacFarlane's bear or the patriarchial bear. It is currently known from a single specimen which was killed on June 24, 1864, by two Eskimos from the Barren Grounds of northern Canada. It was examined three weeks later by the explorer Roderick MacFarlane, after whom it was named. Its remains are now housed in the Smithsonian Institution. In 1918, American zoologist Prof. C. Hart Merriam gave the yellow-furred creature the scientific name *Vetularctos inopinatus* (meaning "unexpected ancient bear").

Saber-toothed survivors

Yet another animal from the past that may still survive is the African saber-toothed cat. The Zagaoua people of

Dangerous relations
This African golden cat may be related to the legendary Tanzanian mngwa. Despite the many killings attributed to the mngwa, zoologists doubt its existence because it has never been captured or photographed.

Ennedi in northern Chad speak of an animal that they call the mountain tiger. This animal has never been seen by Western zoologists, and many doubt that it exists. The Zagaoua claim that it is larger than a lion, but lacks a tail, has red fur banded with white stripes, long hairs on its feet, and enormous fangs that protrude from its snarling mouth.

Tiger tales

During the 1960's, a French hunting guide in the Ouanda-Djailé area of the Central African Republic heard a terrifying roar that his African game tracker identified as that of the ferocious mountain tiger. When the game tracker was shown a series of pictures depicting

> **Local hunters say that the irkuiem resembles an enormous polar bear and estimate that it is five feet high at the shoulders and weighs 1½ tons.**

living and extinct animals, he pointed to the African saber-tooth and stated that this was the mountain tiger.

Like the tigelboat, the Papuan devil pig, and the irkuiem, this big cat seemed to resemble a creature that has not walked the earth for thousands of years. Yet the people of Chad swear that it exists. If such creatures really do exist they are extraordinary survivors from another era. If they do not, they may underscore the tenacity of the human memory, for the saber-toothed tiger, the giant short-faced bear, and the ancestors of the tapir were all hunted several thousand years ago.

FROM FOLKLORE TO FACT

Stories of strange animals abound in folklore. Often frightening and bloodthirsty, these mysterious beasts are generally dismissed by scientists as being too fantastic to be real. But in several cases, these tales have been found to refer to real animals, and real, but previously unknown, animal behavior.

FOLKLORE BELIEFS about animals are often rejected either because the species described is unknown or because the behavior attributed to it does not mesh with current scientific knowledge.

One example is the thunder horse, an enormous creature that the Oglala Sioux of Nebraska, Wyoming, and South Dakota, claim leaps down from the skies to the earth during electrical storms. It was dismissed as a mere flight of fancy until 1875 when the Sioux Indians showed its large bones to paleontologist Prof. Othniel C. Marsh of Yale University's Peabody Museum. Professor Marsh concluded that the bones were the fossilized remains of a rhino-like relative of the horse that roamed North America about 35 million years ago. He called the creature *Brontotherium* (which means "thunder beast").

Bloodthirsty moths

For many years zoologists believed that moths could not bite, because they sipped nectar through a long, coiled tube called a proboscis. But in April 1967 Swiss insect expert Dr. Hans Bänziger began to research the eating habits of a Malaysian moth called *Calyptra eustrigata*, often found on large mammals, which the Jakun people of Malaysia claim can bite.

Since some moths are known to drink mammalian urine, sweat, tears, and even blood flowing from an open wound, Dr. Bänziger wondered if the Jakun belief arose from the sight of the moth feeding on the fluids of animals. He therefore cut one of his own fingers and offered it to one of the moths, wondering if it would lap up the trickle of blood, as some other species of moth do. Instead, the moth stabbed its unusually short, sturdy proboscis into the wound and began sucking up blood, proving that the Jakun account of the moth's habits was

correct. Far from being an innocent lapper of blood from an existing wound, this species pierces its victim's skin and sucks its blood. *Calyptra eustrigata* is the only skin-piercing, blood-sucking moth known to science.

Hibernating birds

Birds such as swallows and nightjars, which appear in temperate countries each spring and disappear again each autumn, were once believed to spend the winter months buried at the bottom of ponds, like frogs, in a torpid state resembling hibernation. But modern science revealed that they actually migrated to warmer climes.

The Navajo Indians were therefore doubted when they claimed that the poorwill, a close relative of the nightjars and the whippoorwill, slept during the winter. Scientists were certain that, like other birds, they did not hibernate, but instead migrated.

But during the winter of 1946 biologist Dr. Edmund C. Jaeger of Riverside City College in California discovered a sleeping poorwill huddled against the rocky wall of a canyon in southeastern California's Chuckwalla Mountains. It appeared to be comatose. The bird's body temperature was 40°F (23°C) lower than normal and its heartbeat was almost imperceptible. Jaeger banded the bird and studied it for several years. He found that it hibernated for an average of 88 days a year. Additional research later revealed that all poorwills are able to enter a torpid state during winter.

Cases like these have convinced many zoologists and other scientists that hunting and farming peoples, who depend for their very existence on their knowledge of the natural world, can often provide useful and accurate accounts of little-known animal behaviors.

Sleeping at will
The poorwill feeds on insects by night and is a master of hibernation. It is able to enter a torpid state in order to conserve energy during the winter when food is scarce.

CHAPTER TWO

MYSTERIOUS SENSES

Some animals are capable of sensory feats far beyond those that human beings can achieve. The secrets of these creatures' abilities are the specially refined super senses they use to survive the challenges of existence.

Many animals have highly developed faculties of hearing, sight, smell, taste, touch, and possibly another more mysterious sixth sense: an awareness of the earth's magnetic field. These super capabilities can reveal aspects of the world that human beings, with less keen senses, can only imagine. In absolute darkness, we cannot detect other warm-blooded life forms, but some snakes can. Certain plants that seem pale and plain to us appear colorful and striated to honeybees, because they can sense ultraviolet pigmentation. Only lately have we learned of these nonhuman powers of perception, and there are still others that continue to bewilder us. Certain animals, for example, are reported to be able to foretell earthquakes, changes in the weather, and even death.

The mystery of migration

Animals make use of such super senses on migratory journeys that may extend for thousands of miles. The longest of these is undertaken by the Arctic tern. For during the course of a year, the tern flies from the North Pole to Antarctica and back again — a round trip of more than 25,000 miles. Scientists still do not completely understand how creatures as diverse as microscopic bacteria and giant baleen whales make these lengthy and extraordinary journeys.

But it is the precision as well as the length of these journeys that astounds scientists. Limpets, for example, forage on seashore rocks at night and move no farther than a few inches from their daytime resting places, but come the dawn, each shell returns to exactly the same spot. Life's journey takes some animals much farther — many salmon return unerringly to spawn in the rivers where they were born.

How do animals follow routes that they have never traveled before and how do those moving during the hours of darkness or in the gloom of the deep sea know in which direction they should go?

But how can a bird confronted with a featureless desert or a wide tract of ocean find its way? The answers seem to be in the sky, and these clues are available both day and night.

Nectar-powered
The rufous hummingbird flies from Alaska to Mexico and back again each year. It times its journey so that it can stop to feed on the way exactly when certain nectar-rich flowers are blooming.

Scientists have studied the migratory habits of birds in an effort to find the answers. In a number of imaginative experiments, pigeons have been fitted with magnets that distort their perception of the earth's magnetic field; starlings have been held in cages with movable mirrors that can change the apparent position of the sun; and indigo buntings have been kept in darkened planetaria beneath starlike pinpoints of light on a ceiling representing the sky. The birds were then monitored to try to establish how much these measures had affected their navigational systems.

Massed monarchs
Monarch butterflies, having emerged from chrysalises around the Great Lakes, flutter to so-called butterfly trees in Florida, Texas, and California, where they congregate during the winter. The butterflies travel as many as 80 miles each day and their entire journey takes them nearly 2,000 miles.

Birds have also been studied in their natural habitats. Night-flying birds have been tracked by military radar or photographed as they pass across the brightly shining disc of the moon; day fliers have been counted by amateur naturalists; and small migrants have been trapped, ringed, released, and caught again and again, the vital number on a band on the leg logged, recorded, and the resulting data analyzed. What all these tests have revealed is that birds do not have a single system, but use several navigational aids and backup systems to find their way around the globe.

Frequent fliers

For example, it is known that many birds migrate along regular flyways, piloting from one landmark to the next. They navigate by remembering the geography, such as the positioning of mountains, valleys, rivers, and lakes. They may also hear and remember the low-frequency "sound signatures" of waves crashing on the shores or of wind blowing through canyons. Other sounds, like the seasonal calling of choruses of frogs or toads, may act as beacons for night fliers.

But how can a bird confronted with a featureless desert or a wide tract of ocean find its way? The answers seem to be in the sky, and these clues are available both day and night. By day, the sun is the most obvious feature, although to navigate by it, an animal must be able to calibrate its position against an unchanging reference source, which is most likely to be the earth's magnetic field.

The mysterious sixth sense that animals seem to have may simply be an awareness of this field. It is possible that birds may be able to detect the field through the minute particles of magnetite, a magnetic mineral, that are coupled either to stretch-sensitive spindles of muscles in their necks or to nerve endings in the connective tissue around the brain. Bees have magnetite particles linked to nerves in the abdomen, while dolphins have disc-like sheets of magnetite magnets in the brain. Even human beings have been found to have particles of the mineral embedded in the tissue of their noses, giving us some awareness of this sense.

The sixth sense

Some suggest that changes in the force of the magnetic field cause miniscule movements of the particles, which are sensed by the nerves. Others claim that the earth's magnetic field induces certain chemical reactions, and that organisms, whether large or small, might be able to respond chemically to irregularities in the field or changes in its orientation.

Mass movement
Huge flocks of snow geese migrate from their summer homes near the Arctic Circle in Siberia, North America, and Greenland, to winter homes as far away as Mexico, China, and Japan.

Green sea turtle
In a miraculous display of marine navigation, Brazilian green turtles swim more than 1,350 miles across the featureless ocean and locate Ascension Island, a tiny dot in the middle of the South Atlantic, before laying their eggs on its beaches.

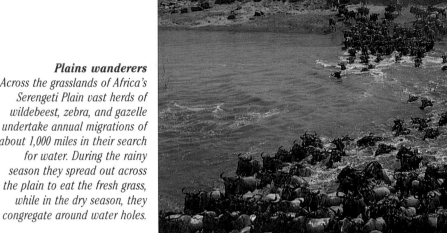

Plains wanderers
Across the grasslands of Africa's Serengeti Plain vast herds of wildebeest, zebra, and gazelle undertake annual migrations of about 1,000 miles in their search for water. During the rainy season they spread out across the plain to eat the fresh grass, while in the dry season, they congregate around water holes.

But whichever theory might be correct, the animal would be able to sense a geomagnetic map of the world. Using this map it could establish its position at any given time and so be able to follow a genetically predetermined compass heading toward its destination.

The internal clock

The flight path of every bird is controlled by its internal biological clock, which detects and responds to daily and seasonal environmental rhythms. This clock is probably set by changes in the length of the day during the year. While older birds may use other navigational tools, a young bird's entire migratory urge and direction is governed by this internal pacemaker, which ensures that it tries to get to the right place at the right time during its migratory journey.

As the bird gets older and learns about its environment, it probably switches from magnetism to other navigational aids. The position of the sun, moon, and stars ensures

Buffalo charm
This green quartzite effigy was made by the Plains Indians to guarantee that the migrating buffalo (a vital food source) would return to be hunted.

that a migrant reaches its destination. By day, for instance, the sun overrides other sensory cues. And if the sun is hidden but there are blue patches in the sky, a bird can use polarized light to guide it through the clouds. At sunset, polarized light from the sun that has dipped below the horizon tells a night-flying migrant the way before it heads off into the darkness.

Night fliers use the pattern of the stars as a nocturnal compass. In the Northern Hemisphere, birds recognize the stars in

---

A bird's migratory urge and direction is governed by an internal pacemaker, which ensures that it tries to get to the right place at the right time.

---

the Pole Star cluster. These are useful since they remain in the same position in the sky because they are above the earth's axis of rotation. Gull chicks have been observed staring at the night sky while sitting in their nests. It is thought that they may be learning star patterns.

Young birds such as geese and swans join family parties and let the adults lead. But cuckoos, which are abandoned in the nests of foster parents, must orient themselves by making solo exploratory flights, which cover increasingly larger areas of countryside around the nest site. But if all these aids fail, the bird can fall back on its genetically determined flight plan and use the geomagnetic field.

Experiments have shown that the migratory behavior of adult and juvenile birds is almost completely different. Starlings, breeding in the Netherlands, for example, fly west and southwest to winter in southern England and northwest France.

Flight program

In one experiment, a batch of adult and juvenile birds was captured, taken to Switzerland, and released. The adults knew they were in the wrong place, adjusted to the difference, and headed northwest to their normal wintering sites. The youngsters, guided only by their inborn geomagnetic sense, flew west or southwest — the direction in which they were genetically programmed to go — and found themselves in the southwest of France and the northwest of Spain.

Feathered stowaway
The wheatear migrates each year from southern Europe to Africa south of the Sahara, but this particular bird is taking a brief rest on a ship during its flight.

The importance of magnetism in bird navigation is made clear when migrants fly in magnetic storms or past magnetic anomalies such as iron-rich mountains or large radar stations. Even on clear days, when the sun is visible and the sun compass should be working, the birds at first become disoriented before they are able to rectify the mistake.

Magnetic pull

The list of living things that detect and respond to the earth's magnetic field is increasing daily. Scientists have discovered that many creatures, including bacteria, algae, primitive flatworms, limpets, termites, honeybees, salmon, tuna, sharks, toads, salamanders, sea turtles, birds, whales, dolphins, rodents, and higher mammals, including humans, can all sense and respond to the powerful pull of the earth's magnetic field.

Coral reproduction
This staghorn coral has just released its eggs and sperm, which float in a cloud above it.

RULED BY THE MOON

At the same hour, on the same night, and at the same time each year, almost every coral along the entire Great Barrier Reef on Australia's Queensland coast releases tiny bundles of eggs or sperm. This amazing synchronization occurs almost exactly a week after the full moon, either in late November or early December, depending on increases in sea temperature. It can be predicted almost to the minute. The first sign is the appearance of pink bulges protruding underneath the mouths of the millions of coral polyps that make up the reef. Just 10 minutes or so later the packets are shot out into the water. It looks like an underwater snowstorm — a blizzard with the snow going upwards.

Lunar cycles

The moon governs the sex lives of many of the creatures and plants in the world's oceans. Among them are polychaete worms, marine sponges, horseshoe crabs, brown seaweeds, and fish. During the spring tides, for instance, the grunion fish of California swarm high onto the beaches in great numbers to deposit eggs and sperm.

The moon exerts its influence on the seas twice each day in the form of tides. Many sea creatures therefore move in time to this 12-hour cycle. The fiddler crab, for instance, becomes very active at low tide and returns to its burrow as high tide approaches.

On land, too, the phases of the moon influence animal behavior. Porcupines, for example, feed for longer periods as the moon wanes. Birds, too, are influenced by the moon. Lapwings, which usually feed by day and roost by night, have nocturnal feeding orgies at full moon and consequently feed less during the day at that time.

The influence of the moon manifests itself in other ways. Antlions, which build pits to trap insects, vary the size of their traps according to the phases of the moon. The volume of the pit is greater when the moon is full. No one understands exactly why — or to what advantage — antlions synchronize their hunting activities in this way with the phases of the moon.

SECRET SCENTS

For most animals, an acute sense of smell is essential for survival. Some animals are so sensitive to odors that they "see" the world as a pattern of scents, just as humans respond to light and colors.

ON MAY 6, 1879, FRENCH NATURALIST Jean Henri Fabre placed a female great peacock moth — the largest European moth — in a jar and left it by an open window in his study. Later that night, Fabre's son came rushing out of the study. "Come quickly," he called in great excitement. "Come and see the moths, big as birds. The room is full of them."

The family rushed into the study to find it swirling with moths, all of them male, each one flying down to the jar and back to the ceiling, time and time again. The same thing happened for several nights afterward, arousing Fabre's curiosity. What was it that attracted the males from far and wide? Several years later and after many more experiments, Fabre concluded that the male moths were attracted to the female by a smell so dilute that the human nose cannot detect it.

Transferring excitement

Eight years later, German scientists discovered the vital chemical that was produced by the female moth. The quantities involved were so minute that they needed the glands of around half a million female moths to produce just 12 milligrams. Scientists named the newly

---

"Come quickly," he called in great excitement. "Come and see the moths, big as birds. The room is full of them."

---

discovered substance "pheromone." Pheromones are, in effect, chemical messengers that transfer excitement from one organism to another.

Subsequent studies have revealed that the female extends glands from her body, releases the scent, and flutters her wings to waft it away. The wind does the rest. The pheromone spreads downwind, and although it is usually very diluted by the time it reaches a passing male, the creature can detect just one molecule if it lands on his antennae. Yet he will probably not change course until he detects about 200 molecules. Then he turns to face the female, and zigzags upwind along her chemical trail. Pheromones from other females do not confuse the male because each female releases her scent in pulses, giving him the rhythmical, directional information he needs to home in on his chosen mate.

HUMAN POWERS OF SMELL

People are often amazed at the incredible sense of smell that many animals seem to possess, and deride the sensitivity of the human nose. But psychologists have found that mothers can identify their children's clothes by their smell alone. And although most of the people who say they are attracted to someone because of his or her smell are referring to the perfume or after-shave that person has applied, some experts now believe that it is the specific pheromone produced by that person that is the real attractant.

Human pheromones

In one famous study, female college students sharing dormitories or apartments found that their menstrual cycles became synchronized. Although this mechanism is not fully understood, it is thought to be due to a pheromone in human perspiration that appears to influence menstrual cycles. Some scientists believe that, like animals, humans can detect these very subtle scents. Human beings, however, detect them only on an unconscious level.

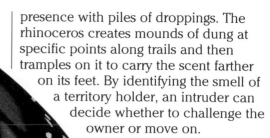

presence with piles of droppings. The rhinoceros creates mounds of dung at specific points along trails and then tramples on it to carry the scent farther on its feet. By identifying the smell of a territory holder, an intruder can decide whether to challenge the owner or move on.

Scent patches
The two black ovals located nearest the body on this orange tiger butterfly's wings produce the creature's powerful scent.

But this system can be exploited by other animals. The Australian bola spider dangles a sticky ball coated with a pheromone very similar to that released by the female moth. Males are attracted to the scent and fly close to the spider. The wing fluttering stimulates the spider to swing the ball, which may trap as many as eight moths. The ball is then hauled up and the victims eaten.

Scent signals

Moths and spiders, however, are by no means the only creatures that rely on scents to find food or mates. Some animals can detect these signals more than three miles away. In some cases, even the merest hint of a scent can have a strong effect. Male tortoises, for example, have been known to become so frenzied when they detect a female's scent that they have tried to mate with a rock or fallen branch over which a female has recently passed, leaving a scent trail.

Many animals use scents to warn other animals to keep away. The territory of a pack of northern timber wolves, for instance, is indicated by a boundary line of urine marks. Rabbits, badgers, and hyenas register their

The nose knows
Chemists evaluate new perfume ingredients in the laboratory. Most humans can discriminate between 6 and 22 smells at high concentrations, but we can be trained to detect more subtle differences. Expert perfumers, for example, can identify more than 10,000 different scents.

Harmonious hives

For social insects such as honeybees, pheromones maintain harmony within a colony; the bees' entire lives are governed by a special cocktail of chemicals produced by the queen. This substance not only serves to attract male bees, it can even prevent other female members of the nest from becoming sexually mature by inhibiting

> ## The substance can even prevent other female members of the nest from becoming sexually mature.

ovarian development. As a colony grows larger, the substance becomes more and more dilute until the queen's influence is reduced and the production of a new queen begins. But there is room for only one queen in a colony, so before a new queen emerges, the old one must leave.

Until recently, this amazing chemical manipulation of the social group was thought to be confined to invertebrates, such as bees and wasps, but a mammalian version is now known. The naked mole rat, a furless, wrinkled, sausage-like animal, lives in colonies below ground in the Somalian deserts of East Africa. The workers are ruled by a queen who maintains her influence by walking over them, distributing her pheromones. Despite cramped conditions, the colonies are peaceful.

Caught in the act
Dogs can be specially trained to sniff out illegal substances such as drugs. Their sense of smell may be as much as a million times sharper than a human's.

ULTRA-VISION

Some creatures possess amazing powers of sight that are uniquely suited to their environment. These powers allow them to "see" things that other animals, and humans, cannot detect.

RED, ORANGE, YELLOW, green, blue, indigo, and violet are the only colors of the light spectrum that humans are able to see — so-called visible light. But there are other forms of light that we cannot see. Infrared radiation is emitted from warm bodies, while ultraviolet light arises from very hot bodies, such as the sun. Some animals have special systems that can detect these forms of radiation, allowing them to "see" things that are not visible to other creatures.

Infrared detectors
Using certain specialized devices, humans can now detect objects that give out infrared radiation. For example, people trapped beneath rubble after an explosion can be located by rescuers using infrared detectors. Pit vipers and rattlesnakes have their own natural infrared detection systems. These creatures have heat-sensitive pits at the front of the snout with which they can sense the infrared "image" of their prey. The system is so sensitive that it can respond to minute changes in temperature many hundreds of times faster than any human-made device. It is a particularly useful sense for rattlesnakes and other predators that search for food in tunnels. The snakes can "see" their victims, even in complete darkness. And in the day, a well-camouflaged animal may escape the eyes of a snake, but it will be given away by the halo of the infrared radiation it gives out.

Ultraviolet guides
Bees, butterflies, and several other insects are able to see features that are only visible in ultraviolet light. This specialized sensitivity allows them to see things in daylight that humans and other animals cannot see.

Seeing in ultraviolet
A human sees the evening primrose flower as a uniform white color (above left). A photograph taken using ultraviolet film (above right) shows how pollinating insects see the flower, revealing the distinct "nectar guides" that lead them directly to the food source.

Flowers, for example, have ultraviolet marks on the petals that, rather like the lights on an airport runway, guide insects to the nectar and pollen at the center. Young flowers or buds, which are not yet ready for pollination, often lack these ultraviolet patterns. They do not attract attention until they need it.

Revealing marks
But this clever system gives rise to another mystery. Predators such as crab spiders can change their color to match the petals of the flowers on which they sit, waiting to attack insects. But in ultraviolet light — the way in which their prey sees them — the spiders are clearly visible, for they seem to lack the ultraviolet marks that are characteristic of the flowers. How they are still able to surprise a victim visiting their flower remains a mystery to researchers.

Several birds are able to detect ultraviolet light, and experts believe that this ability helps them navigate. For example, when the sun is hidden by cloud, ultraviolet light pierces the haze and reveals the position of the sun. Hummingbirds, which are also able to distinguish features in ultraviolet light, are thought to possess color vision that is richer than anything humans can imagine.

Homing in
From a distance this cherry blossom would appear a uniform color to humans, but a bee, which has ultraviolet vision, sees it as having a very dark center, and is guided toward this pollen-rich part of the flower.

BEASTLY FORECASTS

Country folk have always believed that changes in animal behavior signal changes in the weather. But do animals really know things meteorologists don't? And can various animals even predict earthquakes, as the Chinese believe?

During December 1974 the people of Liaoning province in northeastern China began to notice that many of the area's animals were behaving strangely. Geese refused to nest and pigs would not eat; snakes stirred prematurely from hibernation to glide off into the snow, even though they swiftly froze to death; rats roamed the streets, allowing people to pick them up; and normally docile cows fought with one another and attempted to dig into the ground with their hooves. To most observers, these activities would have seemed bizarre and unrelated and so would have been ignored. But the Chinese, with a written history that stretches back 3,000 years, knew

> ## Geese refused to nest and pigs would not eat; snakes stirred prematurely from hibernation; rats roamed the streets; and normally docile cows fought with one another.

from reports in ancient documents that certain changes in animal behavior had often occurred before an earthquake. This gave rise to the belief that animals give advance warning of movements in the earth's crust.

Heeding animal warnings

Four years before the animals of Liaoning began acting strangely, the Chinese authorities had begun to study the likelihood of an earthquake occurring in the province. The pattern of past earthquakes combined with geological studies indicated that an earthquake was imminent. In 1973 the government announced that an earthquake would occur in the next two years.

They trained the local people to detect the early warning signs of an earthquake so that life and property could be safeguarded. Thousands of volunteers were recruited; some were given the task of measuring the earth's activity and monitoring the water supply, but most were instructed to look for strange behavior in animals. From mid-December 1974 such behavior increased dramatically among animals in the wild, who seemed to be aware of tremors before domestic animals. In February 1975 in Fujian province, however,

domestic animals began acting uncharacteristically. Their behavior, when coupled with changes in the groundwater, and minor earth tremors, led the authorities to order the total evacuation of the city of Haicheng on February 4. On that day, the earthquake struck. Although it destroyed half the buildings in the city, most of the 500,000 people who lived there were unharmed.

The Italian experience

More than a year later, in the Italian Alps, animals issued similar warnings, but they went unheeded. On May 6, 1976, the village of San Leopoldo in the region of Friuli was almost completely destroyed in an earthquake that measured 6.5 on the Richter scale.

One of the houses wrecked by the earthquake belonged to the parents of Helmut Tributsch, professor of physical chemistry at Berlin's Free University. When he visited the village after the disaster, neighbors and friends told him that several hours before it struck, many local animals, including dogs, cats, cows, caged birds, and even wild deer, had begun to act strangely.

Tributsch began to collect other reports of earthquake premonition by animals, some dating back to antiquity, and later published them in *When the Snakes Awake* (1982). For example, the mules, cats, and chickens of Liguria, in Italy, were reported to have been extremely agitated prior to a series of terrible earthquakes in January and February 1771 and on May 26, 1831. On February 20, 1835, all the dogs in Talcahuano, near Concepción, Chile, fled from the town, and the sky was filled with frenzied sea gulls — soon afterwards, the town was wrecked by an earthquake.

Howling at the earth

In Messina, Sicily, just before the Calábria earthquake of 1783 that occurred across the Strait of Messina on the Italian mainland, the dogs howled so loudly and hysterically that an order was issued to shoot them. Many were killed, but the rest continued to bark.

A more recent example occurred on February 9, 1971, when two policemen patrolling the streets of a town in the San Fernando Valley in California saw swarms of rats fleeing along the pavements toward the outskirts. A few hours later, tremors shook the valley.

In Japan, where earthquakes are a common occurrence, it is well known that catfish and goldfish in aquariums swim about frantically just before an earthquake. Since water is a good conductor of both

> **Animals are able to hear the minute infrasonic earth tremors that precede earthquakes. Humans cannot detect them.**

A warning ignored
Dogs howled incessantly throughout the night before the San Francisco earthquake, which struck on the morning of April 18, 1906. This photograph of the aftermath shows the central business district, which was almost completely destroyed.

sound and seismic vibrations, it seems possible that the fish may hear or feel the earth trembling. Yet fish in glass containers whose water is insulated from the earth do not appear to "feel" earthquakes. Perhaps, therefore, fish are only able to pick up signals of a chemical or electrical nature.

Infrasonic warning systems

Investigations have led researchers to suspect that animals are able to hear the minute infrasonic earth tremors that precede earthquakes. Humans cannot detect them because such tremors are below the lower limits of our hearing. Snakes and birds, however, are very sensitive to infrasonic vibrations and so are often able to detect the onset of an earthquake. Some species may also be able to perceive electrical changes in the air, which occur when rocks are stressed by the beginning of a quake. Equally remarkable is the ability of many animals to predict changes in the weather. The claims of some farmers that they can foretell meteorological disturbances simply by watching the behavior of their livestock were once dismissed as superstition. But some scientists now believe that these animals too may be responding to changes in the air's static electrical charges that occur before a storm.

Weather-sensitive bees

The capacity of bees to forecast changes in the weather is well understood. Their sensitivity to electrical fields may explain their agitation before thunderstorms and other meteorological phenomena that produce electromagnetic fluctuations.

Sometimes the correlation between animal behavior and weather may be due simply to the amount of humidity in the air. Swallows, for instance, fly higher than normal during dry weather — probably because the movement of the warm, dry air lifts the flying insects on which they prey to greater heights than usual during such weather.

The belief in rural England and North America that seeing dark-colored frogs

Confusing cows
Superstition has it that when cows lie down, rain will soon follow. But since cows often lie down in the sun and stand up just before showers, their forecasts are unreliable at best.

means that rain will soon fall may stem from the fact that during humid periods the granules of dark pigment (melanin) in the frog's skin cells expand, making the frog appear darker; in dry, arid periods, they contract, making it look paler. In many tropical countries, this weather-related color change is so well known that local people often keep frogs in glass cases to act as living barometers.

Patented predictions

In the 19th century, it was alleged that freshwater leeches respond to changes in the weather. One inventor attempted to exploit their weather-predicting abilities by making a leech storm glass. When it was about to rain, he claimed the leeches would rise to the surface of the water so ringing a tiny bell. In sunny weather they would stay at the bottom of the glass.

But the precise mechanisms that cause such changes in behavior and enable animals to forecast the weather remain as mysterious as those that enable them to predict earthquakes.

DISAPPEARING PETS

James Berkland, a former geology professor based in California, says that he can predict when earthquakes will occur by monitoring gravitational readings and geyser activity and analyzing the behavior of cats and dogs.

Earthquake warning

Berkland believes that when large numbers of cats and dogs run away from home, an earthquake is about to occur. To test his theory, he collects reports of lost pets. At the conclusion of his research, if his findings show that a large number of these disappearances occur just before an earthquake, it will lend credence to yet another controversial method of earthquake prediction.

Dr. James Berkland

EXTRAORDINARY ANIMAL POWERS

Animals often have keener hearing, sight, and sense of smell than humans. When they show signs of psychic abilities, these too seem to be stronger than anything humans claim to have experienced.

SOME PETS HAVE SAVED their owners' lives. When Josef Becker of Saarlouis, Germany, took his German shepherd dog, Strulli, into a local inn, the dog became agitated, running around in circles, howling at his master, tugging at his clothes, and trying to drag him from his seat. Strulli caused such a nuisance that Becker put the dog outside.

Somehow the dog got back into the inn and began to tug frantically at Becker's clothes, forcing him to leave the inn. Moments later, to the deafening crash of timber, bricks, and plaster, the inn collapsed on its occupants, killing 9 people and injuring more than 20

> ## "I have seen enough to convince me that animals do have a special psychic power to sense danger before it happens. We are fools to ignore them."
>
> ### Dr. Ute Pleimes

others. Builders excavating next door had damaged the inn's foundations, which then gave way. Some claim that the dog had a premonition that caused his strange behavior, but it seems more likely that his sensitive hearing alerted him to the inn's imminent collapse.

Psychiatrist Dr. Ute Pleimes, of the University of Geissen, Germany, has recorded more than 800 cases of pets warning their owners of impending disaster. "I have seen enough to convince me that animals do have a special psychic power to sense danger before it happens," he said. "We are fools to ignore them."

Harbingers of death

In a phenomenon that some researchers believe is akin to telepathy, certain animals seem to be able to sense the impending death of someone they love.

Capt. A. H. Trapman, in his book *The Dog: Man's Best Friend* (1929), describes how, during the First World War, an Airedale terrier foretold its master's death. The man was a naval officer on minesweeping duty in the North Sea. Whenever he sailed, both his wife and dog

would go to the quay to see him off. The Airedale had never shown any distress on these occasions. Yet, one day, the dog behaved oddly. At the quayside and on board ship, he persistently tried to stop his master from sailing, tugging at his trousers and whimpering. That night the officer was drowned when his ship went down. At about the time the man died, the dog howled inconsolably.

Spirit journey?

English rural tradition has it that the spirit leaves the body just before death, and it is believed that some animals seem to sense this. A cat that belonged to the British prime minister, Sir Winston Churchill, refused to leave Churchill's bed during his final illness. But a few hours before he died, the cat ran away from the house.

Another phenomenon that seems to defy logical explanation is the ability of some family pets to find their owners even when these animals are many miles from home.

> **The dog persistently tried to stop his master from sailing. That night the officer was drowned when his ship went down.**

This occurs so often that some believe that there may be a paranormal link between these pets and their owners.

The most famous such journey was undertaken by a dog named Bobbie who was taken, at the age of two, from his home in Oregon on a long trip. When the car stopped en route in Indiana, Bobbie ran off.

A faithful friend
This statue in Edinburgh serves as a reminder that the link between man and dog can outlast life. Bobby, a Skye terrier, remained faithful to his master even after the man died in 1858. Until Bobby died 14 years later, the little dog kept a vigil at his master's grave in Greyfriars churchyard, Edinburgh, Scotland.

Prison companion
A cat belonging to the Earl of Southampton, Henry Wriothesley, was so loyal that it climbed down a chimney to join him in the Tower of London in 1600.

The family hunted for him but could not find him and had to resume their trip. Three months later Bobbie miraculously reappeared at their house in Oregon. The author, Charles Alexander, authenticated Bobbie's story by placing newspaper advertisements along the dog's supposed route. He then interviewed dozens of people who claimed to have seen Bobbie on his 2,000-mile journey.

Cases of mistaken identity

While Bobbie's case seems genuine, some reports of so-called homing pets may really be cases of mistaken identity. In some cases, however, reportedly unmistakable physical marks identify the animal with certainty. No one can explain how these pets find their way, often traveling hundreds of miles across unfamiliar and sometimes difficult terrain to be reunited with their owners.

A home-loving cat
After being lost in Wales, Samson the cat arrived home two years later — 250 miles away in London. His owners were convinced that the cat was Samson because it looked exactly like him and seemed to know them.

A WINGED RESCUER

Looking up at the sky, she saw a sea gull hovering overhead. The elderly lady took a chance that it was indeed the same bird and shouted: "For God's sake, Nancy, get help."

WHILE WALKING ALONE near her home in Cape Cod, Massachusetts, in 1980, 82-year-old Rachel Flynn accidentally fell over a 30-foot cliff onto a deserted beach. Trapped between boulders, shocked, and too badly hurt to move, she felt sure that she would die before rescuers could come to her aid.

Looking up at the sky, she saw a sea gull hovering overhead. Miss Flynn thought that she recognized the bird. It looked like the one that she and her sister fed regularly at their home and that they called Nancy. She took a chance that it was indeed the same bird and shouted: "For God's sake, Nancy, get help." It seemed like a long shot, but it was the only hope she had.

Tapping at the windowpane

A mile away, Miss Flynn's sister June was working in the kitchen, when she was interrupted by the gull tapping the windowpane with its beak and flapping its wings frantically. Later June recalled that the bird was "making more noise than a wild turkey." For more than 15 minutes, the bird kept up its onslaught until finally, irritated by the noise, June tried to shoo it away. But Nancy would not leave. At last, June decided that, unlikely though it seemed, the bird was trying to tell her something. She hurried from the house to see what the bird wanted.

Flying to the rescue

As soon as June stepped outside, the bird flew off. June somehow knew that the bird wanted her to follow. And indeed the gull flew for only a short distance at a time, stopping frequently to see that June was behind her. Finally, Nancy alighted at the edge of a cliff. When June looked over the cliff, she saw that her sister was lying at the bottom. Nancy had done what Rachel Flynn had asked her to do and brought help.

Later the two sisters puzzled over the rescue. Was it just a curious coincidence — and a good story? Or was it something more? And why would a wild bird help rescue a human being? And how did the bird know what to do?

ANIMAL LANGUAGE

For years, people have been fascinated by the weird sounds produced by certain animals. Some scientists have been convinced by their research studies that certain species use these sounds to communicate with one another.

POSITIVE IDENTIFICATION

It seems that animals with the power of speech can even help to find lost property. In 1985 Marlène Naud's home in Angoulême, western France, was burgled. Among the many items stolen was her prized parrot, Coco. The burglars themselves were never found, but three years later, thanks to her dog, Black, Mme. Naud was reunited with her bird.

Mme. Naud was walking with Black through La Rochelle, a city about 60 miles to the northwest of Angoulême, when she spotted

Marlène Naud with Black and Coco

Coco in the window of a pet shop. She tried to convince the owner of the shop that Coco was her parrot and that it had been stolen three years previously. But the shop owner refused to part with the bird and called the police.

Perceptive parrot

Finally, it was the parrot itself that managed to persuade the police that it belonged to Mme. Naud. The parrot, it seems, recognized Mme. Naud's dog and in front of everybody called out the dog's name. Amazed, the pet shop owner handed Coco back to its rightful owner.

WHEN THE U.S. NAVY used underwater listening devices called hydrophones to detect enemy ships during the Second World War, they were amazed to hear a series of low-frequency sounds that could not possibly have come from any vessel. These strange noises can be heard in all the oceans of the world. Each sound occurs in pulses, and each pulse has a frequency close to 20 Hz (20 cycles per second). Researchers once thought that the loud sounds were caused by the surf pounding on the shore. But studies revealed that the noises increased in intensity in late afternoon and reached a peak at around midnight. Waves, on the other hand, remain regular. The sounds are heard for about 15 minutes, followed by $2^1/_2$ minutes of silence. This pattern provided a vital clue to their origin — it is very similar to the breathing cycle of a slow-cruising whale.

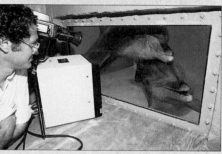

Wailing whales

Soon suspicions were confirmed. Thousands of hydrophones were placed on the ocean floor and everything that passed nearby and every sound was recorded, scrutinized, and identified. Eventually, a fin whale passed over the equipment, and the source of the strange sounds was found. The question now was, why did the whale produce these sounds?

It is known that the toothed whales (which include dolphins and killer whales) travel in groups and constantly call to one another in order to keep their group together. The 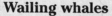 calls vary from group to group, and the dialects are often so distinct that researchers can tell them apart. In some cases, individuals have their own special calls, recognized as a series of whistles and clicks. These creatures seemingly identify one another from their calls, and marine biologists suspect that they may even be exchanging other kinds of information.

Dolphin conversations
Scientists from the Kewalo Basin Marine Mammal Laboratory at the University of Hawaii claim that dolphins respond to their verbal commands, suggesting that the animals understand human language. The scientists "talk" to the dolphins by means of a monitor placed near their tank; a camera enables them to see the animals' responses.

Haunting melodies
The sounds produced by young male humpback whales during the mating season are like songs. They sometimes sing for up to 22 hours.

A voice that carries
Birds that inhabit open spaces, such as the yellowhammer, sing with long, monotonous, constantly repeated phrases that are carried over great distances despite interference from wind, temperature changes, and atmospheric turbulence.

But why should the giant baleen whales, which apparently travel alone, produce sounds that would also appear to be some form of communication? The answer is not yet fully clear, but some researchers suspect that these seemingly solitary whales are actually traveling in herds. These herds may be spread out over hundreds of miles. Perhaps the whales use this loud, low-frequency call because it can be heard across huge distances. Observations of fin whales moving in synchrony but many miles apart indicates that this may be the case.

Sound and survival
The fact that whales can communicate over great distances became known when marine biologists first discovered the remarkable songs of the humpback whale. The long, almost melancholy sounds consist of repeated patterns like those of a bird's song. And they appear to function in the same way, either to space out males (like prairie chickens at their courtship ground) or to attract a mate. The songs are probably learned by baby whales while they are still in the womb, much as chicks first learn to identify and imitate the sounds of their own species while still in the egg.

Sound, it seems, can be vital for a young animal's survival. In many cases, if a mother hen cannot hear her chicks' chirping, she will attack rather than protect her offspring. And crocodile hatchlings call to their mother to encourage her to carry them safely in her jaws from the nest to the water.

In a competitive world in which only the fittest survive, sound continues to be important throughout an animal's life. The songs of birds, for instance, are tailored to their environment: They sing at a frequency and with a pattern that will carry the song the farthest. Birds in forests sing with short, tuneful songs, whose changing frequency is able to pass around trees, cut through foliage, and survive reverberation.

Talking to the animals
Some researchers even attempt to "talk" to their charges. They have taught sign language to chimpanzees, orangutans, and gorillas. And parrots (and even dolphins) are known to mimic the human voice. But it is not clear whether the animals are simply copying their teachers or using the newly acquired language to convey information. And are we interpreting them correctly? Is a gorilla that sees a duck and signs the words "water bird" indicating that it has seen a bird that it associates with water, or is it simply saying it sees water, and a bird? It is a controversial area, but there have been, at least in an interpretive sense, some amazing conversations. To the question: "Where do gorillas go when they die?" one astute gorilla reportedly signed "Comfortable hole, goodbye!"

> If a mother hen cannot hear her chicks' chirping, she will attack rather than protect her offspring.

Keep away
Wolves can deceive their competitors using sound. Their howls reverberate through the forests, giving the impression of a much larger pack, and warning rival packs to keep their distance.

Clever chimps
Scientists at Japan's Kyoto University have demonstrated with the aid of computers that chimpanzees are able to recognize numbers, colors, and objects.

CHAPTER THREE

SECRETS OF SURVIVAL

Certain creatures can survive, and even thrive, in conditions that are so harsh that no human being could withstand them. Exactly how these animals cling to life against all odds is often seemingly inexplicable.

Some animals are able to tolerate or adapt to all kinds of extraordinary conditions, including intense heat and freezing cold. Others can survive long periods of drought. A few animals, particularly insects, can adapt so that they can thrive in poisonous or radioactive environments. Still other creatures, when introduced to new habitats, have been known to take them over, breeding uncontrollably and destroying everything in their way. We may see this as an animal plague, but it is in fact a natural and necessary part of some creatures' life cycles. And a few animals even boost their

chances of surviving by entering into unlikely and apparently inexplicable relationships with other creatures.

Immune to death

All these peculiar types of behavior have one purpose: to ensure the survival of the species. But some of the strangest tales concern those creatures that seem to be capable of cheating death itself.

Many creatures are able to survive conditions that should be fatal. For instance, some bacteria can thrive in nuclear reactors, despite being constantly bombarded with particles of radioactive material. And there is a form of alga that lives in the extremely hot, concentrated sulfuric acid used in certain industrial processes.

While these survival capabilities seem incredible, creatures who can dehydrate almost completely and still survive are even more remarkable for, according to zoological wisdom, life depends on the presence of water. Without water, cell structures collapse and the body dies.

And yet, some animals seem able to survive for long periods of time without water. At one time scientists believed that certain animals — mainly tiny

wormlike creatures called nematodes — actually died when they were dehydrated and that water brought them back to life. It is now recognized that they do not die but instead go into a state of suspended animation from which they are able to revive and continue a normal existence.

Many creatures survive conditions that should be fatal. Some bacteria thrive in nuclear reactors, despite being bombarded with radioactive particles.

In a series of experiments, Prof. Conrad Ellenby of the University of Newcastle, England, showed that nematodes could lose all the water in their bodies and still revive when they were moistened.

The hardy water bear

The ultimate survivor is probably the microscopic tardigrade, or water bear. Water bears are tiny animals with jointed legs closely related to a major group of animals called arthropods. In the wild they live in moss. Scientists found apparently dead water bears in one specimen of moss that had been kept in a museum for 120 years. When the tiny

All dried out

Some types of adult nematodes, tiny, wormlike creatures, can dry out completely for as long as four years, only to revive again when placed in water. The eggs of nematodes are also resistant to drought. It is believed that this ability was developed because some nematodes live in temporary pools of water and have to survive when the pools dry up.

Survivors against all odds

Water bears are tiny, primitive creatures that are able to live for long periods in a dormant state. When dormant, they contract into a ball and dry out. They can remain in this state for months or years and can withstand extremes of temperature and bombardment with radiation. When moistened, they become active again.

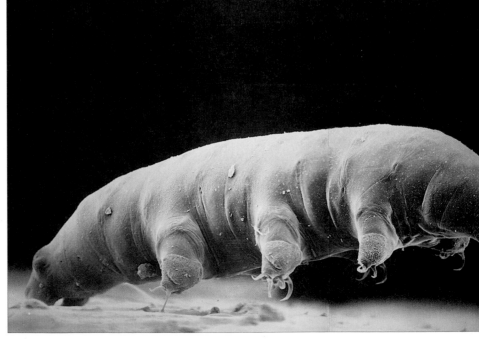

♦ PAGE 66

THRIVING ON POISON

Since the 1940's, insects troublesome to man have been bombarded with killer chemicals. But, in many cases, the insects have not only developed a resistance to these poisons, but actually increased in numbers.

INSECTS ARE AMONG THE TOUGHEST of all animals — especially, it seems, the ones that human beings regard as pests. Untold trillions of mosquitoes, cockroaches, houseflies, lice, and other insects have been killed by insecticides. Yet in any group of insects there are always a few that have been born with a natural immunity to the attacking chemical. These survive and breed. And their offspring simply fill up the ecological "space" left by the dead — so before long there are just as many insects as before, except that they are impervious to existing insecticides.

Adaptive fish
Mosquito fish are one of the few vertebrates to have acquired an immunity to insecticides.

Air spray
Pesticides are easy to apply by air. But since insects can develop a resistance to the chemicals, the dosage must be constantly increased or the chemical changed.

Such insect resistance takes three basic forms. Some insects are born with an ability to use their own body chemistry to make the poison harmless. Others are simply stronger — for instance, they may have a thicker outer protective layer, which keeps the chemical from being absorbed by the body. Others may have a built-in wariness to surfaces or places treated with insecticides, and may simply avoid them.

Acquiring immunity

An insect population as a whole can acquire a resistance to chemicals very quickly. This is aptly demonstrated by certain disastrous experiments conducted with the pesticide known as DDT. In southern Europe, notably in Italy and Greece, both the U.S. military and later the United Nations sprayed vast quantities of DDT in an attempt to destroy malarial mosquitoes. This practice began in Italy in 1945; as early as 1947, the mosquitoes were effectively resisting the poison. As the doses were increased (to the enormous detriment of other wildlife), the mosquitoes acquired early-warning systems and by 1952 had developed an innate behavior that made them avoid places that had been sprayed.

The irony of the modern passion for utilizing such insecticides does not end there. The substances that have been used over the last half-century or so to destroy insects — DDT, lindane, dieldrin, and other related chemicals — do not break down easily into harmless substances, nor do they poison only insects. They contaminate everything they touch, and over a period of time accumulate in the body fat of all types of animals.

Passing on the poison

This means that they build up in the food chain, for as one animal eats another it absorbs the poison in its prey, and because this accumulates in body fat, it cannot get rid of it. Human beings, who are at the top of the food chain, are therefore among the world's heaviest consumers of insecticides, even though the food they eat may not itself have been in direct contact with the poison. Flies and mosquitoes, however, with a reproductive cycle much shorter than that of man, are able to develop resistance to insecticides more quickly.

Magnified louse
Lice have become resistant to most insecticides used against them, so it is now more difficult to eradicate them.

animals were moistened, they revived, if only for a few minutes. If a water bear never entered such a dormant state, its natural life span would probably last about a year. But some will live (albeit somewhat intermittently) for 60 years, due to their ability to go into suspended animation when the conditions do not suit.

Hiding from life

Once it enters what zoologists call a state of cryptobiosis or "hidden life," a water bear becomes virtually indestructible, and can put up with treatment that would kill it (and any other animal) instantly were it in its active state. In the 1920's, P. G. Rahm of Germany's University of Freiberg placed inert water bears in temperatures over 300°F (149°C) and chilled them to below minus 330°F (−201°C) without affecting their ability to revive and survive. At the University of Paris, Henri Becquerel froze water bears as near to absolute zero (−459°F or −273°C) as his equipment would permit. Absolute zero is the lowest possible temperature and reaching it is a physical impossibility. Becquerel was able to take the animals to a fraction of a degree above absolute zero. In theory, no living thing should be able to withstand such cold, but the water bears did.

Similarly, more than 50 percent of any group of human beings will die if they are exposed to a level of 500 roentgens of X-rays over a 24-hour period. Yet it takes 570,000 roentgens of X-radiation to kill half a group of water bears within a day. Extremes of pressure, too, leave the animals intact. Tardigrades can survive being placed in a virtual vacuum. Such conditions would make ordinary terrestrial animals explode — which is why space suits are worn by astronauts outside their vehicles and why high-flying airplanes are pressurized.

Water bears have also survived after having been immersed in carbonic acid, liquid hydrogen, nitrogen, helium, and hydrogen sulfide, any one of which

> **True hibernators curl up in warm nests for the winter, breathe only three times a minute, and will not awaken even if picked up.**

would kill a "normal" animal. Through it all they steadfastly refuse to die. These tiny creatures seem to thrive by breaking the rules of science. Their anatomy and ability to desiccate are their strongest survival features, but scientists are still baffled as to how they activate them.

Sleeping through winter

One of the more well-known — yet still extraordinary — ways in which some animals cheat death is to hibernate, or

Super survivor
Cockroaches are among nature's greatest survivors. They can eat almost anything, are resistant to many pesticides, and may even be able to withstand high doses of radiation. Scientists suspect that if a major atomic explosion occurred, cockroaches would be the creatures most likely to survive.

Sleeping squirrels
Although squirrels, like this white-tailed antelope ground squirrel from the American southwest, hibernate, they awaken periodically to eat the food that they stored in their nests during the autumn.

Hanging around
This lesser horseshoe bat is hibernating. A bat's body temperature drops during the winter, protecting it from some harmful parasites that cannot live at such low temperatures.

and parasites, which cannot survive the low internal temperature. In experiments, when ground squirrels were exposed to radiation, hibernating squirrels were found to suffer fewer ill effects than their active counterparts.

Scientists still do not know how hibernating animals regulate their body temperatures during the winter months. It seems likely that hormones may be responsible. During hibernation, the heart slows down to a fraction of its normal rate and pumps chilled, thickened blood around the body, while the barely perceptible breathing fuels the blood with just enough oxygen to maintain life.

sleep through winter. Hibernation helps these creatures survive because they do not have to endure freezing temperatures or compete with other creatures when food becomes scarce.

Only a few relatively small mammals (and one bird) genuinely hibernate. Some Australian mammals, such as the koala and the spiny anteater, enter a short-term state of torpor, in which their body temperatures become very low, but they are still able to move around. Other animals, like bears, who are often said to hibernate, are merely in a very deep sleep. Their body systems have not shut down to the extent of true hibernators such as squirrels, hedgehogs, dormice, and hamsters.

In cold storage
Hibernation is not strictly sleep: it is more like a state of suspended animation or dormancy. Instead of maintaining their normal body temperatures of about 100°F (37.8°C), hibernating animals keep their temperature just above that of their surroundings, rather as cold-blooded creatures do. True hibernators curl

Sleepy bear
In winter, the koala bear cools off and slows down but does not hibernate.

up in warm nests for the winter, breathe only about three times a minute, and will not awaken even if picked up.

In addition to offering protection from the cold, hibernation confers some other benefits such as defense against disease

Wake-up call
How hibernators actually manage to remain alive while they turn their backs on winter is mysterious enough. But how they are able to wake up at the beginning of spring is one of the unsolved puzzles of nature. With so little energy being consumed, nearly all of the animals' vital functions close down — including, apparently, the activity of the glands that shut them down to begin with. Yet even in the deepest part of their winter sleep, hibernating animals periodically wake up and move about. These waking periods occur more often at the beginning and end of hibernation. Possibly, a type of master biological clock somewhere in the animal's biochemistry tells them how long they have been hibernating and keeps time with the rhythms of the earth. But how they know they should remain asleep longer when spring is late is still unknown.

Death-defying toads
Dormant toads have been found in stones broken open in quarries, as shown in this Victorian book illustration. The cold-blooded creatures generally hopped away from their stony prisons, never to be seen again. If, as some claim, these toads had entered the rock during the geological period in which it was laid down, they might be several millennia old.

ANIMAL PLAGUES

Why do certain animals periodically mass in such huge numbers that they become a veritable plague, sometimes destroying everything in their path so that they eventually starve to death?

THE FIRST SIGN OF AN APPROACHING SWARM of locusts is a smoky cloud on the horizon. As the swarm comes nearer, the sky darkens, the sun is hidden, and the air is filled with the deafening noise of millions of wings. After it passes, the land is left bare of every green plant, for the voracious creatures will have eaten everything in their path before moving on.

Devastation on the wing
The desert locust is the most destructive insect in the world. This grasshopper inhabits the dry and semi-arid regions of northern Africa, the Middle East, Pakistan, and northwest India. Since 1910, there have been six widespread plagues of desert locusts. The area subject to invasions by the desert locust covers approximately 11,600,000 square miles, and a single medium-sized

> ## The desert locust is the most destructive insect in the world. In Ethiopia, a huge concentration destroyed enough grain in six weeks to feed a million people for a year.

swarm may contain up to 1 billion insects, each eating its own weight of vegetation every day. In Ethiopia in 1958, a huge concentration of locusts destroyed enough grain in six weeks to feed a million people for a year. Enormous swarms of up to 250 billion, weighing some 500,000 tons, have been recorded.

Maintaining the balance of nature
Yet these vast, destructive swarms seem to form the means by which the delicate balance of nature is maintained and overgrown populations are regulated.

The potential rates of population increase of all animals are extremely high. Even the slowest breeders, such as elephants, would soon reach fantastic numbers if a large percentage were not killed off, in some way or another, long before death from old age intervened.

The claim has been made that, if all were to survive, the progeny of a single pair of aphids would, within a season, produce a heap of insects as high as Mount Everest. In reality this could never happen because no animal exists independently of its environment.

Greedy grasshoppers
The migratory locust (above), which despite its name is no more likely to migrate than any other locust, is one of eight notoriously destructive species of locust living in the tropical and subtropical regions of the world. Wherever they are found, all species of locusts are feared for their ability to form swarms and destroy crops.

A plague of locusts
Two people encounter locusts swarming in a forested area of Kenya, East Africa.

Escaped killers
A scientist studies a dangerous strain of bees that was produced by East African and American species crossbreeding in the wild.

KILLER BEES

Apart from carriers of disease, the most dangerous insects in the world are bees, whose stings can be lethal to human beings.

In 1956, Prof. Warwick Kerr of the Ribeirão Prêto School of Medicine near São Paulo, Brazil, imported 70 queen bees from what is now Tanzania, East Africa, in an experiment to improve honey production. The following year, 26 of these escaped with their swarms and mated with local honeybees. The result was the notorious killer bee, whose populations began to move northward, toward the United States, at an alarming rate, and have now reached Texas. They have already stung more than 200 people to death.

Predators, such as the tiny larvae of ladybugs, and parasites take their toll, but even if this were not so, food shortages would cut down the numbers long before there were enough aphids to form even a small molehill.

Arctic cycles

Some tropical species, such as the locusts, assume plague proportions at fairly regular intervals; but plagues tend to be even more regular among Arctic animals. This is because the harsh polar environment is simpler: many animals have only one food source and one main predator. There is a 4-year cycle in lemming populations in Arctic regions and a 10-year fluctuation in those of Canada's snowshoe hare (also called the varying hare). These cycles are believed

---

It is known that mass migrations of lemmings occur, during which most of the migrants perish.

---

to be regulated by the time taken before a given phase of the moon appears on the same date of the solar year. As the numbers of lemmings, hares, and, farther south, of voles, increase and decrease, the numbers of their predators such as arctic foxes and snowy owls fluctuate along with them. When numbers reach a peak, two factors begin to reduce the population: food shortages and stress caused by the rigors of extreme overcrowding.

Leaping lemmings

It is known that mass migrations of lemmings occur, during which most of the migrants perish due to starvation, drowning, or the depredations of predators. In 1868 in Trondheimsfjord, Norway, a steamer reportedly sailed for 15 minutes through a swarm of swimming and drowning lemmings two or

three miles wide and very much longer. This, and other similar observations, are probably responsible for the legend that, when their populations increase beyond certain limits, lemmings become frenzied and aggressive and commit suicide in huge groups by hurling themselves into the waters of the ocean.

Searching for greener pastures

But the real reasons behind lemming migrations are much simpler. The furry rodents migrate in search of food when overpopulation leads to overgrazing. When they reach a habitat that does not provide them with suitable food, the lemmings then push irresistibly across the rough mountainsides, tumbling down

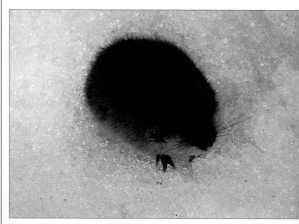

Menacing migrant
Normally timid vegetarians, lemmings become fearless during their periodic mass migrations. They utter doglike barks and may even bite people.

slopes and falling into holes regardless of whether or not they can climb out again. Water is no obstacle and lemmings have been known to swim across fjords more than 2 ½ miles wide. It is not surprising, therefore, that they should sometimes be swept away by the tide; but there is no truth in the idea that they march in army formation to the coast and deliberately hurl themselves into the sea.

Population explosion
The population of the snowshoe hare has a 10-year cycle which may be related to the lunar cycle or the availability of food.

INVADERS FROM AFAR

Man has accidentally or deliberately introduced many species of animals into areas of the world distant from their native habitat, often with quite extraordinary and unforeseen results.

SOMETIMES A WITNESS can be so amazed by the sight of a strange animal that he or she may not consider the possibility that it is a perfectly normal, though non-indigenous, creature. Thus legends build up about a mythical beast. One spring morning in 1883, a woman went to draw some water from Eagle Creek, near Solomonsville, Arizona. When she began to scream, a second woman came running. To her horror, she saw a huge red monster with what seemed to be a man on its back, trampling the woman into the ground. After the monster fled, its hoofprints were found to be cloven. The mysterious beast became known as the Red Ghost, and all manner of bloodcurdling deeds were attributed to it.

When the Red Ghost was finally shot, it was found to be a type of large reddish camel, and the mysterious manlike object on its back turned out to be the remnant of a human skeleton, strapped tightly in place with leather thongs. Presumably the camel was one of many that had been imported into the United States prior to the Civil War. The identity of its ghoulish rider, however, remains a mystery.

Beetle mania

But most animal imports consist of more than one animal, so the creatures are able to breed. And without natural enemies, they can sometimes multiply with alarming speed. For example, in the summer of 1917, about a dozen strange beetles were noticed in a nursery garden in Philadelphia, Pennsylvania. They were identified as *Popilla japonica*, and had probably arrived in 1911 on a consignment of irises or azaleas from Japan, where they are seldom a pest. In America, however, free from natural enemies, the numbers of these beetles increased rapidly. Within seven years, the area they inhabited increased from 3 to 7, then 48, 103, 270, 733, and finally to 2,442 square miles. By 1941, the Japanese beetles covered over 20,000 square miles. Half

Cane toad
With no natural enemies, the cane toad swiftly colonized Australia after it was introduced there in 1935 from South America. This large species of toad is poisonous, and weighs up to five pounds.

the major agricultural pests in the United States, such as the wheat-eating Hessian fly and the codling moth, whose larvae feed on apples, have come from abroad.

Rabbit reduction

Not all agricultural pests are insects. Rabbits too can cause immense damage to crops. Until the arrival of the Europeans, rabbits were not found in Australia, but when they were released there, their numbers grew rapidly. Not until the arrival of the myxomatosis virus was the rabbit population reduced. This virus had previously existed among the cottontail rabbits of Brazil. When introduced into rabbit populations that had no natural or acquired immunity, epidemics invariably followed and rabbit numbers dropped. But, as always happens, a few animals were able to survive the viral attack, because they possessed genes resistant to the virus, and their offspring eventually replaced the nonresistant rabbits. But by the time the population had become immune, its numbers had been significantly reduced.

The cane toad is a native of South America that was introduced into northern Australia in 1935 to destroy a beetle that was devastating the sugarcane crop. But the toads became a greater problem than the beetle — by multiplying so rapidly that their numbers soon attained plague proportions. Moving ever southward, by 1989 the toads reached Brisbane, the capital of Queensland.

Pet toads

Scientists attempted to control the population by luring female toads into plastic bags with the aid of tape recordings of the males' rattling mating cries. But their efforts have been largely thwarted by the cane toad's status as a pet among Australian children. They can be kept as pets because, although poisonous — each toad can secrete enough poison to kill a dog — the poison is only secreted if the creature is attacked.

INEXPLICABLE INTERACTIONS

Throughout the natural world, it is common to find animals of different species forming beneficial partnerships. But some animal alliances are truly mystifying, for they appear to ignore the basic principles of the natural world.

*I*N THE OCEANS OF THE WORLD, a partnership exists that seems to flout the laws of survival The poisonous Portuguese man-of-war floats on the ocean surface and trails a curtain of long tentacles through the water. These tentacles bear lethal stinging cells called nematocysts, which will paralyze any animal that brushes against them. Yet there is one exception — a small striped fish known as the horse mackerel, which spends its life among the man-of-war's deadly tentacles.

The great protector

The tiny horse mackerel appears to be immune to the poison of the Portuguese man-of-war's nematocysts, and in turn it is protected by them from attack by other sea creatures. But what benefit does the man-of-war obtain from this strange partnership? Some scientists suggest that the horse mackerel may act as living bait, and that potential predators attempting to catch the fish are stung by the man-of-war and then devoured. This theory, however, has never been confirmed.

Equally remarkable are the cases in which young animals are reared by females of a different species, especially when the offspring would normally be attacked by their foster mother's species. There have

---

A small striped fish called the horse mackerel spends its life among the man-of-war's deadly tentacles.

---

also been reports of cases in which the foster mother belongs to a species normally preyed upon by that of her "adopted" offspring — as with hens and ducks that have mothered kittens until they were weaned.

No less bizarre are instances in which an animal of one species attempts to mate with an animal of a markedly different species. One example is a 3-ton bull elephant seal called Humphrey, who, in October 1987, attempted to make friends with a herd of dairy cattle on the Coromandel Peninsula of New Zealand's North

FAMILIAR SOUNDS

One seemingly inexplicable, and very one-sided, interaction between different species involves honeybees and one of the world's most striking insects — Europe's deathshead hawk moth, so named because it bears a macabre skull-like mark on its thorax. The moth is often found inside beehives, yet it seems to attract little attention from the bees, so that it can

Impressive impersonator
The deathshead hawk moth makes a loud squeak that seems to protect it from attack.

consume their honey without putting itself in any danger.

Scientists suggest that this situation may be due to its strange ability to squeak loudly. The purpose of this peculiar talent has never been satisfactorily explained, but experts have recently proposed that it may be an extraordinary example of voice impersonation. They suggest that the moth's squeak may resemble sounds emitted by the hive's queen bee. If this is indeed the case, the worker bees are fooled into believing that the moth is a queen bee and thus do not attack.

Mother love
Some animals display bizarre maternal instincts, "adopting" and caring for baby animals of different species. Experts still do not know exactly why this happens.

Island. Abandoning his normal marine habitat, Humphrey chased the cows over a period of five weeks by lumbering up into their pasture. The seal's attentions upset the cows so much that they

---

An operation to find the cause of her stomach pains revealed the presence of three live water snakes.

---

stopped producing milk. It was not until their owner erected an electric fence that Humphrey abandoned his amorous advances and returned to the sea.

Another example of unrequited love involved a young elk that spent 76 days during the winter of 1986 pursuing a Hereford cow known as Jessica, which lived on the farm of Larry Carrara at Montpelier, Vermont. The elk's antics attracted more than 75,000 visitors to the farm. The curious pursuit ended when the elk shed his antlers, and with them, his interest in the cow.

The beast within

It is well known that tapeworms and other parasites can infest the human digestive tract. However, if all the stories are to be believed, this parasitic behavior also extends to some truly strange creatures. One unusual example is that of a Delaware farmer, Clarence Thompson, who, one hot day in June 1908, drank water from a stream and apparently felt a wriggling in his throat. He thought nothing of this until, a few days later, he began to suffer severe stomach pains. These went

A scientific wonder
Nikita, the result of the unlikely mating between a miniature Shetland pony and a wild zebra, plays with her proud father at the South African farm where she was born. Displaying characteristics from both species, Nikita is the only foal born as a result of such a liaison — although zebras and horses have successfully mated.

A friendly kiss
Shirkhal, a fully grown tiger in Toulon, southern France, shows his affection to his "mother." The springer spaniel bitch suckled and reared Shirkhal and his sister, Kali, after they were abandoned by their natural mother at birth.

on for nearly a year until, in June 1909, he reportedly vomited up a live, 3-inch-long lizard-like beast, known locally as the ground puppy (because its snout and head resembles that of a small dog).

More recently, 11-year-old Doreen Luckett swallowed some water while swimming near her home in Darrow in the Mississippi Delta during the summer of 1984. Two weeks later, she was admitted to hospital and the astonished surgeons removed a living, 9-inch-long garter snake from her stomach. In a

Let's be friends
Abandoning its normal instinct to attack, a polar bear plays in the snow with a sled dog.

similar case, in 1979 a Turkish girl named Yeter Yildirim hit the headlines when an operation to find the cause of her severe stomach pains reportedly revealed the presence of three live water snakes. Her doctors surgically removed the creatures and concluded that she had swallowed river water containing snake eggs.

Events like these may well provide a clue to the origin of the water-wolf belief in Yorkshire, England. According to local folklore, swallowing river water can lead to the growth of a frog-like or lizard-like beast called the water-wolf inside the stomach. It was claimed that to get rid of this distressing creature, the patient could either cough it up or kill it *in situ* by eating seaweed or mixtures of methylated spirits, sulfur, and saltpeter (all of which would probably be as lethal to the host as the water-wolf).

RUNNING WILD

Britain may be facing a feline invasion. Already a number of large non-indigenous cats, including pumas, lynxes, and possibly even black panthers, have been reported to be roaming the British countryside.

THE INCONGRUOUS SIGHT of big cats running wild in the fields of Britain may soon become more common. Many of these cats have escaped or deliberately been released from captivity by owners no longer able or willing to look after them — but there is no accepted explanation for the existence of others.

Each year, there are more reported sightings of such cats: In 1984, Trevor Beer allegedly saw several wildcat-like felids on Exmoor, Devon. And puma-like cats have been reported in such locations as Surrey, Midlothian, and the New Forest, to name a few. Edward Noble of Cannich, in Scotland, actually captured an adult female puma, after livestock had been disappearing from his farm for more than a year. Whether this puma was responsible for the deaths of his animals, or whether she was a pet recently released into the wild, has never been definitively established.

Possible offspring?

It is possible that such wildcat specimens may breed in the wild in Britain, but only if a male of one species meets up with a female of its own species — because interbreeding between pumas, lynxes, panthers, and other large cat species rarely occurs; and even on those odd occasions when it does, the offspring are almost invariably sterile.

But there are some very significant exceptions to this. In July 1988, a long-limbed, male specimen of the Asian jungle (swamp) cat, *Felis chaus*, a large, lynx-like species, was found dead at Hayling Island, Hampshire; and in February 1989, another one, also male, was found dead by farmer Norman Evans, close to his farm in Ludlow, Shropshire. Both had died after being hit by passing cars. The jungle cat is native to the Middle East and Asia and is uncommon in British zoos, so it is presumed that these cats had escaped from private collections. Jungle cats do not need to meet up with others of their own kind in order to mate and reproduce because the jungle cat is so closely related to the smaller domestic cat that matings between these two species are known to yield fertile hybrid offspring.

Jungle genes

British zoologist Dr. Karl P. N. Shuker has studied the problem of these cats. He believes that if the Hayling Island or the Ludlow jungle cat mated with domestic cats who had run wild (known as feral cats) before being killed, there may now be cats living in the Hampshire and Shropshire country-side that possess jungle cat genes. Since fertile hybrids are often very fit and of notably large size (a phenomenon called hybrid vigor), successive generations of breeding between such hybrids and other feral cats could ultimately mean that Britain may soon be home to a new breed of big cat. Since these big cats would be as large as their jungle cat progenitor, it is a chilling prospect.

The Kellas cat found in northern Scotland is another example of

Jungle cat
This fierce-looking felid, known as the Asian jungle or swamp cat, was found dead near Ludlow, Shropshire, in February 1989.

such a domestic/wildcat hybrid. First spotted in 1984, this mystery cat is predominantly black in color but flecked all over with long, white primary guard hairs. Its broad head contains jaws that display striking canine teeth. Over the last eight years, several Kellas cats have been captured or shot, and extensive research and genetic tests utilizing tissue samples have been conducted on them. As predicted by Dr. Shuker in his book *Mystery Cats of the World* (1989), these tests have revealed the Kellas cat to be a complex domestic/Scottish wildcat hybrid.

TEMPERATURE EXTREMES

Some animals appear to defy the laws of nature by living in temperatures so hostile that a human being would perish in moments. How do such amazing creatures survive the extreme conditions to which they are exposed?

*A*CREATURE THAT CAN TOLERATE a temperature of 219°F (104°C) for more than an hour and survive immersion in liquid helium — which, with a temperature nearing absolute zero (− 459°F or − 273°C), is used as a freezing agent in laboratories — may seem incredible. But such a creature does exist. In certain areas of West Africa, larvae of the midge *Polypedilum vanderplanki* inhabit the pools that form in the hollows of unshaded rocks during the rainy season. These hollows fill and dry out many times each year, but the larvae survive these extreme changes without harm. How do they do this?

Natural protection

In the 1950's and 1960's, Prof. Howard Hinton of Bristol University in England carried out various studies on these amazing creatures. He discovered that the larvae survive because, when water is unavailable, they enter a state of almost complete desiccation. In this state, they are able to survive exposure to extremely high and low temperatures. Strangely, when water becomes available again, they absorb it and are able to return to their normal state. They have the extraordinary ability to do this over and over again.

Perhaps Prof. Hinton's most fascinating discovery was that if he cut a desiccated larva in half, both halves, when placed in water, absorbed moisture and wriggled for several seconds before finally "dying." This happened even when the larva had been maintained in the dehydrated state for many years.

Many insects and some animals have developed a remarkable mechanism for surviving unfavorable conditions; they dramatically reduce the rate of their bodily functions. This inactive state is known as diapause. Depending on the species, diapause may occur at any stage in the life cycle, but with insects, it occurs most commonly in the egg. In diapause, the eggs of some insects are resistant to heat, cold, and drought. Larvae of the mosquito *Aedes vexans*, for example, cannot survive without water, so the eggs

Termite mounds
In Australia, the Amitermes *species of termite builds huge, vertically flattened mounds that can rise up to six feet tall. The structures are oriented to keep the mound at as close to an even temperature as possible. The sides, facing east and west, face toward the sun in the early morning and evening, when it is cooler. And at midday, when it is extremely hot, the narrow edges face the sun.*

Nightly excursions
Jerboas, which inhabit the deserts of North Africa and Saudi Arabia, avoid the sweltering daytime sun by hiding underground and emerging from their holes only at night to forage for food.

Summer slumber
Tadpole shrimps, which inhabit rain pools in North Africa, survive the hot, dry summer in an inactive state known as diapause. In this state, they can survive temperatures as high as 217°F (103°C) for up to 16 hours.

remain in diapause until the soil on which they are laid is flooded to form a pool suitable for the larvae.

Getting into hot water

Even when active, some animals can apparently withstand surprisingly high or low temperatures. But in most cases, the conditions are not as extreme as they first appear. Although the temperature of the water of hot springs almost reaches boiling point (212°F or 100°C), few of the insects and fish that live in them can tolerate more than 104–122°F (40–50°C) for long. The answer to this paradox is that the creatures lurk near the surface of the water and at the edges of the spring, where the water is much cooler.

A similar explanation applies to many desert animals, which survive only by managing to avoid exposure to extremes

> **Mice, in their quest for food, have been known to make homes in the huge industrial refrigerators in which meat and other chilled foods are stored.**

of temperature during the day. These creatures, which include spiders, insects, scorpions, and rodents, emerge from their retreats and burrows only during the night. Many desert animals have also adapted their diet. Kangaroo rats, for example, do not need to drink water, and can exist on a diet of dry seeds.

Built for comfort

The huge, rock-like structures that can be seen on many desert landscapes may look like geological features, but they are in fact elaborate cooling systems built by thousands of termites to survive the hot, tropical conditions in which they live. In the deserts of Australia, the mounds of *Amitermes* are built with their narrow ends pointing to the north and south and their sides facing east and west. The sides with the larger surface areas thus face toward the sun in the

Well insulated
This Canadian baby harp seal is well insulated from the cold by a layer of fat below its skin and a thick fur coat.

morning and evening, which warms the nest when the air temperature is fairly low. But at midday, when it is very hot, a much smaller area faces the sun.

Left out in the cold

At the other extreme, there are animals that must avoid freezing to death. Many of the adaptations to a cold environment are external. Birds and mammals of polar regions, for example, possess thick fur or feathers that insulate them against the cold atmosphere. In addition, they have stout bodies and short limbs, proportions that reduce their surface area. But some mammals have even become adapted to unnatural environments. Mice, in their quest for food, have been known to make homes in the huge industrial refrigerators in which meat and other chilled foods are stored. They survive because they develop abnormally long, dense fur, and since they have unlimited food, they can sustain the high rate of internal activity required to prevent a dangerous drop in body temperature.

Other creatures have fascinating mechanisms to keep themselves from freezing. For example, the Alaskan blackfish *Dallia pectoralis* survives the freezing temperatures of the surrounding water by producing an alcoholic substance that prevents its blood from freezing. Contrary to popular myth, however, this fish cannot emerge alive after having been frozen for months in blocks of ice.

KEEPING COOL

Found in the semi-desert regions of eastern Africa, the naked mole rat is one of the most curious of the rodent family. Not only does it live in a society that is unique among the mammals (like termites and other social insects it lives in huge colonies governed by a queen), but it spends its entire life underground. This rodent's extensive burrow systems are well insulated by the heavy clay soil above, so that it rarely experiences any significant temperature increase despite the heat outside.

The naked mole rat lives on roots and tubers, which it finds in its tunnel systems, so that it does not have to expose itself to the hot, dry conditions to forage for food. The animal's strange, naked appearance is probably also a means of keeping cool — it has no fur to trap heat.

Naked mole rats

CHAPTER FOUR

ENIGMAS OF EXTINCTION

Some of nature's most intriguing mysteries involve the plants and animals of the distant past. Although they have left behind many clues, we are still unsure exactly how they lived or why they disappeared.

Through the study of fossils, we now know that the millions of different species that are living today form only a fraction of the different animals and plants that have ever lived. Most species, in fact, have vanished, and we do not know why they disappeared. Fossils range from the skeletons of huge dinosaurs to traces of tiny plants and animals that can be seen only with a microscope. Most fossils are formed from the hard parts of plants and animals, such as seeds, shells, bones, eggs, and teeth. Yet even leaves and pollen have provided clues to the vast array of past life. In order for an

THE GEOLOGICAL TIMESCALE

The earth was formed from a cloud of dust 4,500 million (m.) years ago. As it cooled a crust formed, and gas and water vapor formed an atmosphere of thick cloud and poisonous gas. By 3,000 m. years ago the planet was stable enough for the first living organisms, simple bacteria, to appear. The next 2,400 m. years were dominated by simple forms of life. Complex organisms appeared only 600 m. years ago. Dinosaurs appeared around 225 m. years ago, and humans around 4 m. years ago. Each twist of the spiral below covers 570 m. years of the earth's history.

⑬ **Tertiary period**
65 to 2 m. years ago. Great spread of mammals.

⑫ **Cretaceous period**
145 to 65 m. years ago. Flowering plants evolve; dinosaurs eventually die out.

⑪ **Jurassic period**
208 to 145 m. years ago. Dinosaurs dominate the land.

⑩ **Triassic period**
245 to 208 m. years ago. Reptiles, including dinosaurs, evolve.

⑨ **Permian period**
290 to 245 m. years ago. Mammallike reptiles take over the land.

⑧ **Carboniferous period**
363 to 290 m. years ago. First reptiles appear.

⑦ **Devonian period**
409 to 363 m. years ago. Sharks colonize the seas; insects and amphibians appear.

⑥ **Silurian period**
439 to 409 m. years ago. First plants and air-breathing animals appear.

⑤ **Ordovician period**
510 to 439 m. years ago. Reef-building corals develop.

④ **Cambrian period**
570 to 510 m. years ago. First fish, jellyfish, and mollusks appear.

③ **2,500 m. years ago**
Widespread volcanic activity.

② **3,000 m. years ago**
First bacteria appear.

① **4,500 m. years ago**
Earth's crust begins to form.

⑭ **Quaternary period**
2 m. years ago to the present date. Today, humans dominate the land.

object to be fossilized, it must be buried quickly before it decomposes. Because of this, only a very small proportion of the succession of different plants and animals have survived as fossils.

Even though the fossil record is not complete, it does give us a useful picture of the earth's history. Prehistoric life falls into stages, known as periods, each of which lasted for many millions of years. Fossils inform us of the various forms of life that existed during each time span. But many questions remain unanswered. For example, despite the countless fossils of dinosaurs around the world, we still do not know whether these creatures, which dominated the land for more than 160 million years, were warm-blooded or cold-blooded. Neither do we know why they suddenly died out.

Mass extinction

The extinction of a species, like the appearance of a new one; is a natural part of evolution. It has been happening since life began on earth, and it has claimed nearly 99 percent of all species that have ever lived. Many died out when they were unable to compete with better-equipped species. Some died because of slight environmental changes. But studies of fossils have shown that the story of life on earth has been punctuated by mass extinctions in which many different species of plants and animals suddenly died out together. Exactly how these mysterious episodes occurred, and why certain animals and plants survived the catastrophes, remains a matter of great debate. Some people believe that mass extinctions have been triggered by brief, cataclysmic events, such as the impact of a comet or periods of intense volcanic activity. Others argue that they were brought on by dramatic changes in the climate or sudden outbreaks of disease.

End of an era

The fossil record suggests that five major extinctions have occurred thus far, each marking the end of a certain period of prehistoric life: these are the Ordovician (439 million years ago), Devonian (363 million years ago), Permian (245 million years ago), Triassic (208 million years ago), and Cretaceous (65 million years ago). Each extinction removed a number

Armored animals
Trilobites, so named because they were divided into three distinct parts, or lobes, became extinct at the end of the Permian period. Their bodies contained a hard mineral that was resistant to decay, making them among the most commonly found fossils.

of animal and plant groups that characterized their particular period. For example, a large number of marine families including the nautiloids (sea-dwelling mollusks) vanished at the end of the Ordovician period, the hard-bodied trilobites disappeared at the end of the Permian period, and the ammonites, well known from their coiled circular shells, finally died out along with the dinosaurs at the end of the Cretaceous period 65 million years ago.

The world's greatest extinction
The Permian extinction, 245 million years ago, was the greatest and most mysterious extinction of all time. There are no craters and no great deposits of volcanic lavas, yet this extinction was so massive that roughly 96 percent of all species living at that time were wiped out. It was the first extinction to affect life on land significantly. During previous extinctions most life had been confined to the water. But by the end of the Permian, long after the land had been colonized by amphibians, plants, and insects, reptiles, in their turn, had become well established, and mammal-like reptiles ruled the land. Nearly all of these creatures died out. Nor was marine

The story of life on earth has been punctuated by mass extinctions in which many different species of plants and animals suddenly died out together.

life immune. Many of the marine animals managed to survive until the crisis was over, but the trilobites, rugose corals, and some spiny fish vanished forever.

Some scientists believe that the planet itself was the culprit. For the world was changing drastically during the Permian period. All the continents were moving together to fuse into a single great landmass, known as Pangaea, which stretched almost from the North Pole to the South Pole. At these extremities, Pangaea and its adjoining seas, were too cold for many animals. At the same time, the land grew drier.

Burial at sea
With a single continent now in existence, the amount of warm, shallow, offshore water, one of the richest habitats, shrank. The winters became extremely cold, and a series of crippling ice ages caused the sea levels to drop. Since there was less water, sea salts became more concentrated. Some scientists suggest that this highly concentrated salt water sank to the bottom of the oceans, leaving water above that was too fresh for many forms of marine life, including the corals and tiny crustaceans. This condition might also account for the enormous deposits of salts found in rocks formed during the Permian period.

Other experts question this theory. Studies of the effects caused by a drop in sea level today show that only a small proportion of marine families might

Curious creature
These fossilized remains belonged to an amphibian known as a branchiosaur that disappeared at the end of the Permian period. The strange skull structure of this creature represents the larval stage of an amphibian which seems never to have developed a fully adult skeleton.

become extinct — a number nowhere near the figures lost during the Permian extinction. Also, the massive falls in sea level of hundreds of feet during the last Ice Age (one of the most extreme ever) did not appear to affect life so dramatically.

Cosmic radiation?

Some scientists suggest that the destructive force may have come from outer space. They claim that the cyclical rotation of the galaxy could have placed the earth in a position to receive high levels of cosmic radiation at the end of the Permian period. It is known that such radiation can cause genetic mutations in organisms, and supporters of the cosmic radiation theory suggest that this occurred on a mass scale at the end of the Permian period, disrupting life cycles and ultimately causing extinction. Yet it is unlikely that such extraterrestrial forces would have been so devastating to marine life, which was, at least, sheltered by the sea.

Populations poisoned?

Perhaps the most implausible extinction theory that has been put forward so far is that marine life was poisoned during the Permian extinction, affecting life farther up the food chain. Modern examples of sea pollution are well known, but the Permian poisoning must have been very different to kill off whole animal families worldwide. The chemical substances phosphate, vanadium, and potassium are found in great deposits in rocks dating from the Permian period, and some scientists claim that these may have been released faster than marine animals could deal with them.

Fossil evidence
A fossil leaf of the seed fern Glossopteris *from New South Wales, Australia. The present distribution of these fossils indicates that there was an ancient Southern Hemisphere "supercontinent" that began to fragment about 190 million years ago.*

Yet few experts support the idea that high levels of toxic chemicals could have spread through the seas to such an extent and with such an effect on life on land.

Most scientists explain the Permian extinction by emphasizing three factors: colder temperatures, less salty sea water,

---

If the Permian extinction had not taken place, dinosaurs, mammals, and, ultimately, humans, might never have developed.

---

and lower sea levels. Each of these alone would have upset the environment, but together they might have caused the widespread disaster that befell life both on land and in the sea.

A fresh start

The Permian extinction — like other extinctions — was massively destructive. Yet the wide-scale disappearance of groups of animals and plants created fresh opportunities for those that remained. If the Permian extinction had not taken place, creatures such as the dinosaurs, mammals, and, ultimately, humans, might never have developed.

Shelled survivors
This fossil ammonite, found in Dorset, England, has a characteristic coiled, chambered shell. Sheltered by the sea, the ammonites survived the Permian extinction but disappeared 65 million years ago during the period when the dinosaurs died out.

THE BURGESS SHALE

The bizarre fossils preserved in an area of compressed shale in Canada provide a unique glimpse of an animal community of 530 million years ago — one that was perhaps more varied than our own.

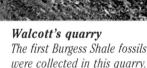

ERY OCCASIONALLY, fossils of soft tissue, which usually decay during fossilization, are found. In 1909 the accidental discovery of an immense array of beautifully preserved and delicate fossils (many of them soft-tissue fossils) in a bed of compressed shale in British Columbia, Canada, caused a furor among paleontologists. For these fossils are all that remain of one of the earliest animal communities of the Cambrian period (570 to 510 million years ago). The fossils have puzzled scientists for more than 80 years.

animals do not fit into any group known today. These unique fossils include strange forms of jellyfish and worms. In most of them, the fine structural details of the animals are well preserved, and in some cases, even the internal organization can be easily seen.

Immense variety

What is so intriguing about the Burgess Shale community is its complexity and diversity. Here, perhaps, is evidence that the natural world does not necessarily become more diverse with time. In some ways, the animal world represented by the Burgess Shale community was more varied than the animal world today. In the Cambrian world of the Burgess Shale, animals existed with anatomical designs that have never been replaced in the subsequent 530 million years. In fact, the animals showed such a vast range of body shapes and styles that if we were to include them in the animal kingdom today, we would have to create 20 new groups in order to accommodate them.

Many of these creatures appear to have died out without leaving any descendants. Extinction of 20 or so of these odd types of animals meant fewer opportunities for later diversification. If extinction had, at this early stage in life, removed a different set of animals, the story of subsequent life on earth would have been very different.

The worm that walked
This fossilized wormlike creature, Hallucigenia, appears to possess seven pairs of legs and seven tentacles. It is so bizarre that scientists are not sure whether it merely represents part of a larger animal.

A unique community

Preserved as shadowy films of minerals, the Burgess Shale community (named after a 19th-century governor of Canada) contains many different species of both soft-bodied and thinly shelled creatures. They belonged to a deep-water world, living on the sea floor or swimming just above. Most of the hard-bodied animals, including trilobites and sponges, can be fitted into the classification of known animals, but many of the soft-bodied

Walcott's quarry
The first Burgess Shale fossils were collected in this quarry. It is named after the discoverer of the fossils, Charles D. Walcott, and is situated in Yoho National Park, near Banff, Alberta.

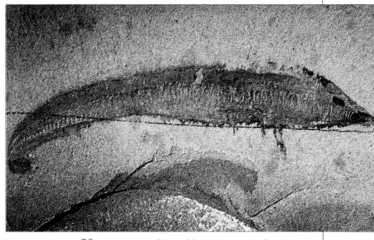

Our oldest ancestor?
Fossils of Pikaia, a flat, fishlike animal, were found in the Burgess Shale. It was stiffened by a rod-like backbone, so experts believe that Pikaia was the earliest chordate, the group of backboned animals to which humans belong. If this is true, this strange creature may be our oldest ancestor.

Fiery cataclysm
About 65 million years ago, the earth suffered what may have been its most cataclysmic event. Whole populations, including all the dinosaurs, died out in one massive extinction. Some believe this mass death was caused by a large meteorite.

DEATH OF THE DINOSAURS

The age of the dinosaurs came to an end 65 million years ago. But it was not just the dinosaurs who died out: In a mass extinction still puzzling scientists today, many other animal groups were destroyed.

*E*VIDENCE IS GROWING that about 65 million years ago, when dinosaurs ruled the earth, meteorites colliding with the planet caused the most famous extinction ever known. Global winds, fires, tidal waves, and dust clouds resulted, and the earth was plunged into a miserable period of darkness and destruction.

Anything in the immediate vicinity of a large meteorite impact would have been vaporized. With the sun eclipsed by clouds of dust and smoke, plants and marine photo-plankton at the far corners of the globe would soon have died, deprived of essential solar energy. With the vegetation gone, herbivores would soon have starved to death. The consequence for carnivores further up the food chain would have been equally disastrous. Although they had existed for about 155 million years, the dinosaurs could not survive these cataclysmic changes and so became extinct.

Catalog of destruction

Along with the dinosaurs went other large reptiles. From the seas the aquatic reptiles — the ichthyosaurs and plesiosaurs — and the aquatic lizards — the mosasaurs — were lost. On land, all the vertebrates

Although they had existed for about 155 million years, the dinosaurs could not survive these cataclysmic changes and so became extinct.

weighing more than 50 pounds died out. The masters of the air, the flying lizards or pterosaurs, became extinct, and marsupial mammals suffered severe losses. The annihilation did not stop there. Invertebrate animals also suffered. All ammonite species disappeared along with many brachiopods and bivalves (mollusks equipped with two shells). Even many of the tiniest creatures, such as the planktonic foraminifera (single-celled animals with a hard skeleton) became extinct.

This mass death marked the boundary between the Cretaceous and Tertiary periods and is known as the K-T event (K from the Greek *kreta,* meaning "chalk" and T from *Tertiary*). It was a shock to all life on earth.

Continents in motion

Continents constantly shift position on the earth's surface. When dinosaurs first appeared, the continents were just beginning to break away from one super landmass, Pangaea. During the dinosaurs' reign, the continents drifted apart, opening up seaways and causing changes in climate.

Pangaea: one landmass formed about 250 million years ago

Pangaea began to split up over 190 million years ago

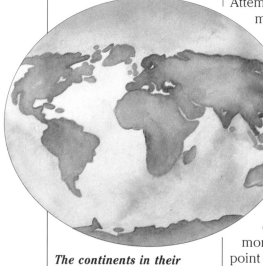

The continents in their present positions

Although it is normal for some individual animal and plant species to die out over time, the families to which the species belong do not become extinct, because some other species within that particular family survive.

Sudden death

Dr. Robert Bakker of the University of Colorado has estimated that, on average, a dinosaur species, such as *Barosaurus lentus* or *Stegosaurus armatus*, might have lasted for 2 to 3 million years. But the K-T event killed off about 15 percent of all animal families. That is as if all pet cats suddenly died out as well as all the lions, tigers, leopards, jaguars, cheetahs, pumas, lynxes, and all other feline animals. Yet the cats are only one family of animals. At the end of the Cretaceous period, every member of many other families also became extinct. The mass death of the dinosaurs and other animals changed the future of all life on earth.

Of course, even this extraordinary number of extinctions does not match the Permian extinction, during which 96 percent of families became extinct.

Gradualism or catastrophe?

Attempts to unravel the puzzle of these mass extinctions have divided scientists into at least two camps. Some scientists believe that during the K-T event the rate of extinctions, usually occurring over thousands or even millions of years, was speeded up by a variety of factors. Others believe that some catastrophic shock, such as a meteorite or meteorites hitting the earth, ended many different life forms in a matter of months or years. But both groups point to evidence of a deterioration in the earth's climate at the end of the

Deadly impact

This artist's impression shows Tyrannosaurus rex, *the fiercest carnivorous dinosaur, watching the huge meteorite that was to seal its fate falling to earth.*

Cretaceous period. For one group it was like the steady onset of winter, while, for the other, it was a sudden storm.

Life for the dinosaurs seems to have been stable for more than 155 million years. Conditions before extinction may have been deteriorating but they were not disastrous. The climate was warm, and there were plenty of flowering trees and subtropical ferns for the vegetarians to feed on and plenty of vegetarians to satisfy the appetites of the meat eaters. If the dinosaurs alone had disappeared, it might be easier to understand what went wrong at the end of the Mesozoic era, but to account for the disappearance of a wide range of animals and plants, both marine and terrestrial, is more difficult.

Dinosaurs on the wane

There is evidence that Cretaceous life was on the wane before the K-T event. Not all dinosaurs died 65 million years ago. Fossil evidence shows that many died long before then. Like other groups of animals, of course, dinosaurs were always dying. Scientists believe that dinosaurs were a well-established and varied group between 5 and 10 million years before the extinctions. But in their last 5 million years, they became more scarce, and there were fewer varieties.

Fossil finds

The fossil evidence in rocks in Montana suggests that from 700,000 to 300,000 years before their extinction the numbers of distinct dinosaur groups fell from 19 to 12 and finally, just before extinction, to 7. As dinosaurs were on the wane, early mammals seem to have been increasing. Some scientists argue that whatever occurred at the end of the Cretaceous period was not a cataclysm but simply the final straw for all the dinosaurs.

Long legs

Impression of the front leg of a dinosaur found in Utah. Dinosaurs differed from all other reptiles in having upright legs.

A few scientists, however, believe that the dinosaurs did not become extinct at the end of the Cretaceous period. While

> ### Some scientists argue that whatever occurred at the end of the Cretaceous period was not a cataclysm but simply the final straw for all the dinosaurs.

they accept that something happened on our planet 65 million years ago to cause the extinction of many groups of animals, some paleontologists believe that dinosaurs were among the survivors.

They claim to have discovered the teeth and some bones of tyrannosaurid, hadrosaurid, and ceratopsid dinosaurs in rocks in Montana deposited soon after the K-T event, which they claim show that dinosaurs lived on for a short time. Most scientists, however, believe that these dinosaur remains were originally deposited at an earlier time and were subsequently eroded out of these rocks and then redeposited at a later stage.

Because the shapes and sizes of leaves tend to be the same in the same climate, fossil plants can be used to identify climatic characteristics of earlier times. Jack Wolfe of the U. S. Geological Survey and Garland Upchurch of the University of Colorado Museum and the National Center for Atmospheric Research in Boulder, Colorado, showed that, during the Late Cretaceous period, the average annual temperature in North America was above 70°F (21°C) with a low rainfall and little seasonal change. This climate extended north as far as the latitude of what is now Portland, Oregon. Just below the present-day Arctic Circle, there were some seasonal changes, with temperatures varying between 58°F and 60°F (14°C and 15°C). Rainfall was sparse. In the far north, temperatures were lower, but not freezing.

Pleasant climate

Wolfe and Garland believe that this pleasant climate varied little until the K-T event, when rainfall increased. Other scientists, however, claim to find a slow and steady deterioration in the weather of the

Chalk white

The white cliffs of Dover in southern England are made up of the chalk skeletons of millions of microscopic algae called coccoliths. The rock deposited at this time is so chalky that the geological period is called the Cretaceous: kreta *means "chalk" in Greek.*

WHAT WERE DINOSAURS?

Dinosaurs first appeared in the Triassic period about 220 million years ago and continued in one form or another until their extinction 155 million years later. They were reptiles, like today's crocodiles, turtles, and lizards, laid eggs, and had scaly skin. But their great advantage over other reptiles lay in their leg structure. Unlike their sprawling relatives, dinosaurs had upright legs to hold their bodies off the ground.

Rulers of the earth

Dinosaurs lived on every continent, including Antarctica. The different species ate a variety of diets. There were carnivores and herbivores, as well as animals that ate insects, shellfish, eggs, and fish, and others, the omnivores, that ate everything.

But dinosaurs lived only on land. A few, such as the hadrosaurs, probably swam in the shallows of rivers and lakes, but the seas were the domain of other reptile groups such as the ichthyosaurs and plesiosaurs. Dinosaurs could not fly. The long-fingered pterosaurs (derived from the same reptile group as the dinosaurs), such as *Dimorphodon* or *Quetzalcoatlus*, with wingspans of up to 36 feet, shared the air with primitive birds, such as *Archaeopteryx*, and flying insects.

▶ PAGE 90

DINOSAURS AT THE MOVIES

Whether as cartoons, real lizards made to seem larger, people dressed in dinosaur suits, or computer-aided models, dinosaurs are more popular than ever on screen.

THE DINOSAUR HAS BEEN featured in films almost since the movies began. Its screen appearances can be divided broadly into two cateogories: the prehistoric, as in such movies as *One Million Years B.C.* and the modern-day, in which dinosaurs are suddenly discovered in some mysterious place where, by a fluke of climate or geography, they have survived unchanged. Such films include *The Lost World* and *The Land that Time Forgot.*

In 1993 the film director Steven Spielberg introduced a third variety of dinosaur movie, one in which computers and other forms of technology play a key role. *Jurassic Park* is about man-made dinosaurs created by scientists and is taken from Michael Crichton's best-selling novel of the same name. It tells of an experiment to develop living dinosaurs by using the DNA extracted from dinosaur bones found by paleontologists. The creatures in question are used to stock a theme park, but things go terribly wrong. The dinosaurs in the film were animated by computer, making them more realistic than the creatures that appeared in earlier movies.

Computerized dinosaur
Computer technology can now be used to create very realistic special effects. This robotic *Allosaurus* forms part of a collection of robotic dinosaurs made by a Californian company. The plates and skin are sculpted and painted to make the dinosaurs appear as authentic as possible. The creatures' joints are operated by compressed air and the movements are carefully controlled by computer.

The Land that Time Forgot (1975)
In 1975 some of the movie world's least convincing prehistoric animals hit the big screen as survivors from a First World War German submarine with prisoners on board landed on a lost island named Caprona. This imaginary island was inhabited by cavemen and dinosaurs. The prehistoric reptiles were played by men in rubber suits, leading to some very unrealistic sequences.

One of Our Dinosaurs is Missing (1975)
The dinosaur in the title is a skeleton in which a secret piece of microfilm is hidden by a gang of Chinese criminals led by Peter Ustinov. The film is set in London in the 1920's, and the action centers around the Natural History Museum.

The Lost World (1960)

Hollywood liked this story so much, they made it twice, once in 1925 and again in 1960. The plot concerns a professor who is sent by a newspaper to investigate rumors of a prehistoric world existing on top of a remote plateau in South America. The dinosaurs in the 1960 version are ordinary lizards specially photographed to appear very much larger than life.

One Million Years B.C. (1969)

In this film, acted without any dialogue at all, two rival tribes, the Rock People and the Shell People, battle for a young woman, played by Raquel Welch. Some of the dinosaur scenes are very convincing. The movie title is less true to life — the last of the dinosaurs died out about 65 million years ago.

At the Earth's Core (1976)

Testing an enormous machine designed to drill through the earth's crust, a scientist discovers a secret, subterranean land inhabited by feuding tribes and dinosaurs.

The Valley of Gwangi (1969)

A Western with a difference, this movie tells of a traveling circus that discovers a secret, dinosaur-inhabited valley in Mexico and captures one of the creatures for exhibition. It features several typical Western duels — including one between an elephant and a dinosaur!

Gertie the Dinosaur (1909)

"She eats, she drinks and breathes! She laughs and cries! Yet she lived millions of years before man inhabited this earth and has never been seen since!" So claimed the advertisements for the adventures of a cartoon dinosaur from the earliest days of cinema. Gertie's creator, Winsor McKay, used the film in his vaudeville act, projected on a screen behind him, making it appear as if he were training the prehistoric creature to do as he commanded.

Father and son team
Dr. Walter Alvarez (left) and his father, Prof. Luis Alvarez, look through a transparent star dome. Together they combined their expertise in geology and physics to create a new theory of how the dinosaurs died out.

Fish lizard
This fossilized Ichthyosaurus *swam in the shallow seas that covered what is now Germany before becoming extinct just before the dinosaurs did. Any theory that attempts to explain why dinosaurs became extinct must also explain the extinction of other groups of animals, which like the ichthyosaurs, became extinct at about the same time.*

Cretaceous period. They believe that the subtropical vegetation was gradually replaced by more temperate woodlands.

Volcanic explanations

Some scientists link these changes and the gradual decrease in the number of dinosaur species with the slow retreat of shallow, mid-continental seas and a period of intense volcanic activity. Somewhere between 68 and 64 million years ago, volcanic lava repeatedly poured out of the earth, covering a huge area of what is now India with lava more

The K-T layer
This layer of clay (marked with a white spot) was laid down between the rocks of the Cretaceous and Tertiary periods at the time the dinosaurs died out.

than half a mile thick. The lava covered over a million square miles. Reasoning from the effects of modern volcanic explosions, scientists theorize that these eruptions might have been followed by dense dust clouds and acid rain.

It is possible that during the later part of the dinosaurs' reign the evolving and shifting earth, through movements of

landmasses, changing sea levels, and volcanic explosions, provoked a gradual climatic change, irreversibly upsetting the balance of all life. Many plants and animals, including the dinosaurs, simply could not adapt to these changes — perhaps because their food sources had disappeared — and so became extinct. Others, such as the birds and mammals, were able to adapt and thrive.

A blow from space

But many scientists reject this theory and believe that the evidence in favor of the meteorite hypothesis is more convincing. The most remarkable evidence lies in a thin layer of clay found at many places in the world, which was laid down in the boundary between the rocks of the

> Movements of landmasses, changing sea levels, and volcanic explosions provoked a climatic change, upsetting the balance of life.

Cretaceous and of the Tertiary periods. This clay band, no more than a few inches thick at most, was first found at Gubbio in Italy by Prof. Isabella Premoli Silva of the University of Milan.

Dr. Walter Alvarez, a geologist from the University of California, studied the clay layer in the 1970's in order to learn just how long the K-T extinction had taken to occur. He and his colleagues came up with a figure of about 1,000 years. Other scientists working on the same project arrived at even shorter times — as little as 50 years — no more than a blink of the eye in geological time and far too sudden a change for most living creatures to adapt to and survive.

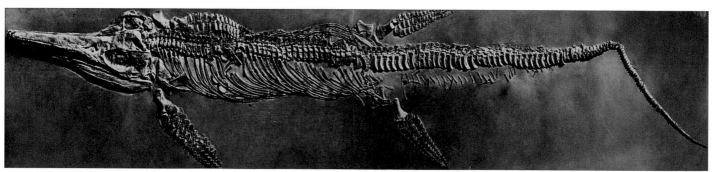

▶ PAGE 92

DEAD ENDS

There seem to be almost as many theories explaining the dinosaurs' extinction as there were dinosaurs. From blindness to boredom, slipped discs to shrinking brains, the list is a long one.

IN THEIR ATTEMPTS TO UNRAVEL the mystery of the dinosaurs' disappearance, scientists have come up with many theories that at first may seem plausible, yet quickly prove to be untenable.

Some of these ideas seem merely foolish. It is not likely, for example, that the dinosaurs died of boredom. With 155 million years of life behind them, why should they be bored? Even if they were, human experience suggests that few have ever really died of boredom. Other claims, for instance that the brains of dinosaurs were shrinking, are easy to disprove from fossil evidence. The brains of the last dinosaurs, such as *Troodon*, were relatively larger, not smaller, than those of earlier dinosaurs.

Nest robbers

Other theories are, at first glance, more plausible. For instance, some claim that small, furry mammals brought down their reptilian neighbors. They did this apparently by changing to an egg diet — dinosaur eggs — eating the dinosaurs to extinction. But it is unlikely that all dinosaurs species in all parts of the world would be affected similarly and simultaneously. Neither is it likely that an animal would cause the extinction of its prey. If dinosaurs became extinct, no more eggs would be produced. How then would the mammals survive?

Some fossils of dinosaur eggs do suggest a problem in a dinosaur species near the end of the Cretaceous period. The eggshells seem to have become thinner with time, to the point where the eggs could not survive. This happens in birds today and is caused by hormone imbalances often brought on by stress. The dinosaurs were supposed to have suffered from the stress of overcrowding, yet the many dinosaur nesting sites in Montana do not show thinned eggshells. Here, the dinosaurs have carefully avoided overcrowding, at least while caring for their young.

Plant plot

Some claim that plants caused the downfall of the dinosaurs. Herbivorous dinosaurs were dependent on the plant food available and they in turn were eaten by the carnivorous dinosaurs. Problems with the plant world would have affected the dinosaur food chain.

Some claim that the evolution of flowering plants, and with them butterflies and moths, spelled an end to the dinosaur world. The caterpillars of these insects are able to devour great swaths of leaves. In the competition for plant food, the herbivorous dinosaurs might have come in second and starved, along with the carnivores. Or perhaps dinosaurs, if they had a poor sense of taste, ate the wrong kinds of these newly arrived plants and died from poisoning.

Unfortunately for all these ideas, the first flowering plants came into existence about 60 million years before the dinosaurs' extinction and no one knows when butterflies and moths became a significant part of the dinosaurs' world. Even if one or two careless dinosaurs choked on a mouthful of flowers, there is no good reason to believe that mass poisoning swept through the dinosaur community.

Thus these theories fail to explain the extinction of the dinosaurs because they ignore all the other animals that died out at the same time, and they tend to overlook the worldwide scale of the destruction. They simply do not explain why some animals and plants died out and others survived.

On the track of extinction
This natural cast of the footprint of a Dilophosaurus *was found in the Painted Desert in Arizona. Theories about the death of the dinosaurs rest on interpretation of fossil evidence such as this.*

Relic of the past?
The idea that dinosaurs are still alive in some remote parts of the world is a persistent one. This illustration shows a plesiosaur, a water-dwelling reptile that lived at the same time as the dinosaurs. It was reportedly spotted alive in Patagonia in 1922.

ROUTE TO EXTINCTION
Many of the marsupial mammals of South America became extinct not because of objects falling from space but because of invaders from the north.

For about 60 million years, North and South America were separate; marsupials flourished in the southern continent, and placental mammals occupied the northern continent.

Crossing the land bridge
Three million years ago, a land bridge formed, linking the two continents, and allowing the animals to move freely between them. Those heading north included the sloths, opossums, armadillos, anteaters, and porcupines, while pumas, foxes, bears, horses, camels, and mastodonts went south.

Not all of these animals could survive in the space available. The placental mammals were better able to compete. As a result 50 percent of the mammal groups living in South America today came from the north, while only 20 percent of those in North America migrated from the south.

Alvarez turned to his Nobel physicist father, Prof. Luis Alvarez, also of the University of California, for help in confirming these figures. Luis Alvarez devised a method of calculating the time it took to deposit the boundary clay by measuring the amount of extra-terrestrial, cosmic particles that had fallen to earth and collected in the clay. Since these cosmic particles fall at a steady rate, he reasoned, the quantity in the clay would be a good indicator of time. He chose to count the amount of iridium in the clay, since this metal element is rare on the earth's crust — there are about 0.03 parts per billion. It is, however, much more abundant in meteorites.

Clues in the clay
When the Alvarez team analyzed the clay, they found so much iridium in it that it seems certain that a whole mass

After living in the shadow of the dinosaurs for millions of years, the mammals had the opportunity to evolve.

of iridium, not just a steady stream of particles, must have fallen at the same time as the dinosaurs' extinction. The most likely cause is a huge meteorite colliding with the earth.

Since that first discovery of a high level of iridium at the K-T boundary, similar high levels have been found in other parts of Europe, in Africa, Asia, North and Central America, and New Zealand, and in sediments beneath the oceans.

The earth holds other clues that lend support to the theory that an object from outer space caused the extinction of the dinosaurs. The clay also contains spherules of once molten rock droplets; grains of quartz with structural signs of shock, like those found at known meteorite impact craters; and the mineral stishovite. The shocked quartz grains could possibly have been formed by volcanic activity, but outside of the clay layers, stishovite occurs only at crater sites. Soot deposits are also found in the clay bands; they are probably a sign of huge global fires that scorched wide tracts of vegetation.

Rapid changes in climate, worldwide extinctions, impact minerals, iridium anomalies, fire deposits, and the other features that mark the catastrophic step from the Cretaceous to the Tertiary period all seem to match the effects that might have been caused by a giant meteorite.

What might have happened? This is one scenario: An asteroid six miles wide traveling at more than 60,000 miles per hour might have collided with earth, creating a crater 90 miles wide. It would also have caused an explosion many times greater than the combined detonation of all the world's nuclear weapons. A terrible winter would have followed, created by the clouds of sooty smoke generated by raging fires, thus killing most life forms on earth.

Crater site

But for many years, no one knew where such an impact crater might be located. Detractors of the meteorite theory claim that if it were true, the immense crater ought to be visible. But supporters of the theory argue that the meteorite might have fallen into the sea or that the crater might have been destroyed as the margins of drifting continents were subsumed. Recently, however, evidence of a huge crater site has been found in the Caribbean, on Mexico's Yucatán Peninsula, and in Haiti.

The extinctions at the end of the Cretaceous period could, however, have been caused by more than one impact. If it is proved true that a series of meteorites fell on earth over a period of years, it might help to explain the spread in death rates. Impacts from meteorites are not uncommon and there are many identifiable craters.

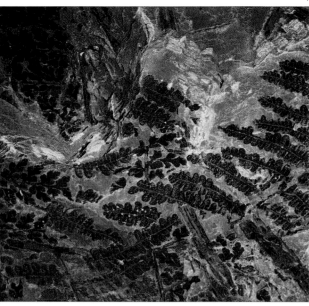

(Prior to satellite photography, however, few of these craters were recognized.) Studies by Eugene Shoemaker of the U.S. Geological Survey have produced statistics for the frequency of different-sized meteorites. Small craters fewer than 6 miles wide occur at 110,000-year intervals. Large 90-mile-wide craters occur, on average, at intervals of 100 million years. David Raup, a statistical paleontologist at the University of Chicago, has matched three of the five massive extinctions in the earth's history with known craters. He has also written a statistical analysis for extinctions that concludes that they may be regular events, occurring every 26 million years.

Survivors of the cataclysm

Many animals, however, survived and even proliferated after the destruction of the dinosaurs and their world. Small placental mammals were not greatly affected. Neither were the fresh-water crocodiles and sea turtles, nor the sharks and bony fishes. Crustaceans, gastropods, and sea urchins felt only ripples while other animals were being swept away.

The birds thrived on the change. It is now believed that birds developed from a group of small, sharp-toothed, bipedal dinosaurs, and that they are probably the dinosaurs' closest living relatives.

So when the dinosaurs left, leaving only fossilized traces, the mammals had room to grow. After living in the shadow of the dinosaurs for millions of years as tiny, cautious, ratlike creatures, they now had the opportunity to evolve into the myriad forms of life that we know today.

Fossilized ferns
In rocks deposited in North America just before the time of extinction (65 million years ago), fossil pollen from flowering plants makes up 70 to 80 percent of the pollen and plant spores. After the extinction, the deposits are dominated by the spores of ferns.

Wolflike predator
An artist's impression shows what Cynognathus, a reptile that was a precursor of the first mammals, would have looked like when it was alive 225 million years ago.

Bird or dinosaur?
Archaeopteryx *used to be classed as the first bird, but now some say birds are just feathered dinosaurs.*

DINOSAUR DISCOVERIES

Dinosaurs have been extinct for more than 65 million years, but new species are constantly being discovered as collectors pry them from their stony hiding places.

WHEN FOSSIL EVIDENCE of a new dinosaur is discovered, it is only the beginning of a time-consuming and complex investigative process. Dinosaur excavation, preparation, and study can take years. For instance, *Minmi*, the first armored dinosaur, from Australia, was found in 1964 but was not removed from its surrounding rock cover until the 1970's, and it was not identified until 1980. *Minmi* lived about 115 million years ago and was protected in life by a coat of bony plates. In recent years many new dinosaur discoveries have been made in Australia.

Both newly discovered and well-known dinosaurs are not always perfect specimens or indeed perfectly understood. It is not unusual for scientists to revise the work of others as new information becomes available, sometimes even changing the dinosaur's name. One of the most famous dinosaurs, *Brontosaurus*, is not really a separate species: in 1903 it was decided that it was a dinosaur called *Apatosaurus*, which had already been described and classified.

Dinosaur without a name
The Kunming dinosaur was about 17 feet tall and lived in south-western China during the Jurassic period, about 200 million years ago. Its 52 sharp teeth show that it was a meat eater. Discovered in 1987, it has not been named and classified, so it is known only as the Kunming dinosaur.

Incredible eggs

Scientific knowledge of young dinosaurs was given a boost in 1979 by the spectacular fossil discovery in Teton County, Montana, of a dinosaur nesting site that shed new light on dinosaurs' early development. The fossilized remains were of *Orodromeus* (meaning "mountain runner") a small, bipedal dinosaur that was a fast-running plant eater, a creature that probably relied on speed to escape from its predators. Nineteen well-preserved eggs were found on a rocky outcrop now known as Egg Island, that once had been a nesting

Model nest
This model shows the young of Orodromeus, *a small dinosaur that walked on its hind legs. It is based on fossilized eggs found at a 75-million-year-old nesting site in Montana in 1979.*

Killer claw
Twelve inches long, the gigantic claw bone of Baryonyx *gave the dinosaur its scientific name, which means "heavy claw." It is believed that only one finger on each hand had such a claw.*

site for *Orodromeus.* The eggs were laid in a spiral, with the outer eggs leaning against a vertical egg in the center.

Cracking the secret of the shells

One of the discoverers of *Orodromeus,* John R. Horner of the Museum of the Rockies, analyzed the eggs using a CAT scanner. The scan revealed that one of the eggs contained a complete fossilized embryo. This tiny creature was revealed

---

Nineteen eggs were found on a rocky outcrop that once had been a nesting site for *Orodromeus.*

---

when the shell was carefully chipped away. The bones of the embryo were well developed, suggesting that the young of *Orodromeus,* like the young of many modern reptiles, were able to fend for themselves as soon as they hatched. The young of other dinosaurs were more like modern birds, which are helpless and must be cared for by their parents.

Killer claws

In 1983 fossilized bones were found in a clay pit in southern England by an amateur fossil collector, Bill Walker. An enormous claw bone, 12 inches long, was the first piece he spotted. The creature to which the bones belonged was classified as a new dinosaur called *Baryonyx,* meaning "heavy claw."

Baryonyx lived about 124 million years ago, and its remains are the most complete of any large, carnivorous dinosaur from the Early Cretaceous period. *Baryonyx* had a long, tooth-lined snout and a narrow skull, which gave an almost crocodilian shape to its head. Scientists believe that *Baryonyx* had a huge claw on one finger of each hand, which may have been used for hooking fish to eat. The full description of *Baryonyx* took scientists at the Museum of Natural History in London over 10 years to complete.

Chinese treasures

In China, references to dinosaurs go back to the Western Jin Dynasty in A.D. 265-317. More recently, dinosaur fossils have been collected and studied scientifically since the 1930's. Now, in cooperation with Canadian teams from the Tyrell Museum of Paleontology in Alberta and the National Museum of Natural Sciences in Ottawa, the Chinese have begun to uncover the secrets of their dinosaurs.

New evidence in the form of dinosaur bones, eggshells and footprints, found in the Gobi Desert, suggests that China's dinosaurs evolved separately, cut off by huge inland seas from Europe and by the ocean from North America. The arid environment they lived in was also very different from the lush and swampy world of the North American dinosaurs.

But despite the many new dinosaurs discovered during the last few years, many more almost certainly remain hidden. Scientists are certain that examples of all the dinosaurs that ever lived have not yet been discovered. It is possible that they never will be.

Fossil quarry
A paleontologist frees fossils from a quarry wall at Dinosaur National Monument on the border of Utah and Colorado.

DIGGING UP DINOSAURS

Paleontologists use various tools in their efforts to extract dinosaur bones and other fossils from the rock in which they are buried. The fossils are cut free from the surrounding rock with drills, hammers, and chisels, so that they can be transferred to a museum for further study.

Fragile fossils

Before being moved, the fossils are often painted with resin to harden them or wrapped in plaster or polyurethane foam so that they are not damaged on their journey. At the museum, the rock is chipped away from the bones. The bones are then painstakingly assembled so that the dinosaur can be studied and classified.

Showing some backbone
The Kunming dinosaur's bones were visible during its excavation from hard, red rock in China in 1987. Two fossilized teeth from a large, carnivorous dinosaur, possibly its killer, were found near its rib cage.

MAMMOTHS

Ten thousand years ago, mammoths roamed the Northern Hemisphere. Their bodies are sometimes still found perfectly preserved in the Siberian wastes.

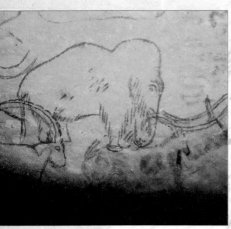

Mammoth art
A Stone Age artist scratched this drawing of a mammoth into the wall of a cave. Primitive man made enormous technological advances toward the end of the Pleistocene epoch about 10,000 years ago. With their stone-tipped spears, they may have upset the balance of animal communities near the end of the Ice Age, perhaps causing the extinction of the mammoth.

WHEN THE LAST ICE AGE was at its maximum extent, about 18,000 years ago, the northern ice sheet covered much of Europe and North America, and ice masses appeared in parts of South America, Australia, and New Zealand. Woolly mammoths, and their elephantlike relatives the mastodons, were an everyday sight for cave-dwelling, Stone Age people, living on the icy margins of the glaciers.

The permafrost of Siberia, which is sometimes frozen nearly a mile deep, holds the carcasses of mammoths that have been preserved in ice for thousands of years. Frozen mammoths have also been found in Alaska, Canada, and Greenland.

Frozen food

Frozen mammoths are often uncovered during thaws, when the northern rivers erode their banks. Sometimes, because the carcasses are so well preserved, the flesh has been eaten by wolves. But, in many cases, even the stomach contents are intact, so scientists can determine the season in which the animals died.

The mammoths that had roamed North and South America, Europe, and Asia for more than 2 million years became extinct about 10,000 years ago, at the end of the last Ice Age. This was probably caused by a variety of factors: Humans may have played a role in their extinction, along with climatic and vegetational changes. Some believe that the extinction of large mammals, such as the mammoths, cave bears, and saber-toothed tigers, was caused by human hunters.

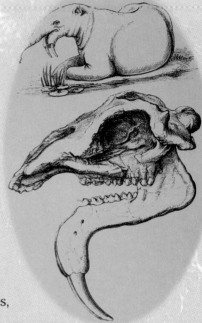

Prehistoric elephant
Deinotherium wandered over Europe, Africa, and Asia more than 2 million years ago. It is unique because it has no upper tusks; its long lower tusks were efficient tools for digging up roots.

Mummified mammoth
This baby mammoth was about six months old when it died 10,000 years ago. It was found when bulldozers were excavating a site in Yakutskaya in Siberia in 1977 and is the best-preserved mammoth ever discovered. The soft tissue at the end of the trunk, usually eaten by scavengers before scientists are able to examine it, was completely preserved, revealing the two fingerlike projections at its end, previously known only from cave paintings. Mammoths probably used these, as elephants do, to pick up food.

Model mammoth
This reconstruction of a mammoth in 1860 exaggerates the actual size of the animal. An adult stood about 10 feet tall.

Mammoth vegetarian
This early 19th-century illustration shows an adult mammoth. To protect against the Ice Age cold, the mammoth had thick, reddish-brown fur. Beneath the fur, it had a $3\frac{1}{2}$-inch layer of fat. Scientists believe that the mammoths used their tusks to clear away the snow so that they could eat the grasses and bushes that lay beneath. One specimen was found with a mouth full of buttercups, frozen as it ate its last meal.

Miocene mammals
Deinotheres and mastodonts flourished in Eurasia and Africa during the Miocene era about 10 million years ago. As in modern elephants, a coat of hair would have been a disadvantage in the hot climates where they lived, so they were virtually hairless.

Ivory trade
This Victorian engraving shows mammoth tusks awaiting collection at the London Docks. Mammoth tusks from Siberia have been traded as ivory for centuries and once provided half the world's supply. It is estimated that 550,000 tons of tusks are still frozen in the permafrost along the north Siberian coast.

LIVING FOSSILS

Amazingly enough, some animals and plants have survived, almost unchanged, for hundreds of millions of years. Why have these creatures thrived while others have become extinct?

*S*TUDIES OF FOSSILS HAVE SHOWN that since life on earth began, the natural world has changed enormously. The continuous process of evolution has determined that most modern species of plants and animals are very different from their fossil ancestors. But there are exceptions to the rule; some animals and plants living today are almost identical to certain fossils hundreds of millions of years old.

For example, along the east coast of the United States, the Gulf of Mexico, and the east coasts of Asia, a relic of the distant past still exists. It is the horseshoe crab, the sole living representative of a once diverse group of fossil invertebrates called *Xiphosurida*. The crab's rounded dome, ranging from brown to dark green, is hinged to a hard, roughly triangular abdomen

---

The horseshoe crab has survived almost unaltered since the first members of the group appeared about 500 million years ago.

---

that ends in a long tail spike. Fossils show that the horseshoe crab has survived almost unaltered since the first members of the group appeared about 500 million years ago. It is now a classic example of a living fossil.

No competition

Another form of marine life that is a living link with ancient creatures is the lampshell, *Lingula*, found in the sandy intertidal area of mud flats and sandbanks. Also known as the tongue shell, *Lingula* is a brachiopod, a form of marine shellfish that was once common throughout the oceans, but suffered heavy losses in the Permian extinction. *Lingula* is one of the most common of brachiopod fossils and, in some parts of the world, whole layers of rock are composed solely of fossils of *Lingula*. Studies of these fossils show that *Lingula's* modern counterparts are almost identical to those that lived about 500 million years ago. Thus, while other groups of brachiopods have come and gone, the lingulid family has persisted unchanged.

These marine animals are now found only in the black, smelly mud of the western Pacific region, particularly off the coasts of Japan, southern Australia, and New Zealand. They live in vertical burrows,

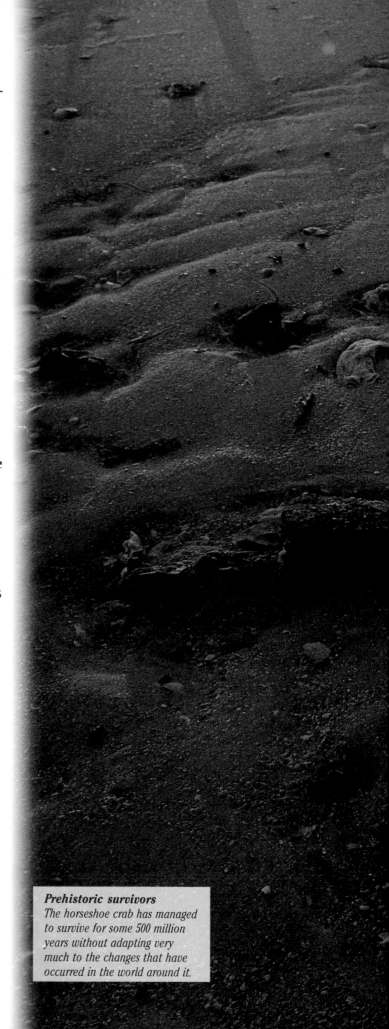

Prehistoric survivors
The horseshoe crab has managed to survive for some 500 million years without adapting very much to the changes that have occurred in the world around it.

UNLIKELY ANCESTORS

Ever since it became apparent that the modern horseshoe crab, *Limulus*, is strikingly similar to the fossil horseshoe crabs that lived hundreds of millions of years ago, scientists have been trying to discover why these creatures should have survived

Fossil horseshoe crab

at all. Experts remain unsure why they exist today, with no modern improvements or specializations, and in competition with highly developed crustaceans such as modern crabs, lobsters, and their relatives. Yet the horseshoe crab is not a true crab. In fact, the group to which the horseshoe crab belongs may have formed the ancestors of the Arachnida, the huge class of animals that comprises the scorpions, spiders, mites, and ticks.

The tuatara is the only true survivor of a group of reptiles that was prominent about 200 million years ago.

anchored by a fleshy pedicle, or foot, that extends deep into the sand. One reason for the creatures' continued survival is their capacity to live in foul water that few other animals can tolerate. In such water they have had few competitors for available food and very few predators.

Coming up for air
Some modern fishes also seem to be prehistoric survivors. Lungfishes first made their appearance nearly 400 million years ago, but 200 million years later, numbers and species were dwindling. Fossil records reveal almost worldwide distribution in ancient times, but modern lungfish are restricted to the rivers and lakes of Africa, Australia, and South America. The living Australian species, *Neoceratodus*, has changed little from the earliest fossil specimens. Lungfish are known for their ability to breathe air through a kind of lung, created and modified from pouches in the gut. The most primitive of the living species, the Australian lungfish, cannot survive drying out, but is able to live in stagnant water by gulping air from the surface. Thus it can tolerate conditions that other fish cannot survive in.

Reptilian relic
Although modern reptiles look similar to their prehistoric forebears, most of them are of quite recent evolutionary origin. However, a large, spiny specimen known as the tuatara is the only true survivor of the Rhynchocephalia, a group of reptiles that was prominent about 200 million years ago. This creature, which is now found only in

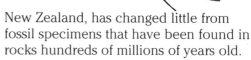

A link to the past
Unlike modern cephalopods (sea-dwelling mollusks), the Nautilus has a chambered external shell, linking it more closely to prehistoric fossil forms such as the ammonites.

New Zealand, has changed little from fossil specimens that have been found in rocks hundreds of millions of years old.

When the tuatara was first described in 1831, it was thought to be yet another newly discovered lizard. But in 1867 it was classified as a rhynchocephalian, mainly because it was found to have a different skull structure from modern lizards and a primitive style of vertebrae. The creature is believed to have survived because it lived in relative isolation on islands where there were few predators.

Prehistoric plants
It is not only animals that have survived the forces of evolution. Some plants have also changed very little since the early prehistoric years. In ornamental parks and gardens across the world, an elegant tree with delicately veined, fan-shaped leaves can often be found. This species, *Ginkgo biloba*, is the only survivor of a once diverse group that first appeared about 200 million years ago. *Ginkgo biloba* naturally grows in the wild only in a small area of China. Its leaves are easily recognizable when fossilized, enabling experts to date the plant to about 60 million years ago. The extinction of all but this last ginkgo species was probably due to competition from the spread of flowering plants.

Airtight chambers
South American and African species of modern lungfish can survive drought by curling up in chambers they dig in the mud of a riverbed as it dries out. Fossil burrows of this type are known, proving that ancient lungfish had the same habit.

Horsetails
Equisetum or the horsetail is the only surviving sphenophyte, a group of primitive plants that flourished about 300 million years ago.

REVERSING EXTINCTION

Some of the world's extinct animals have been brought back to life by a seemingly miraculous series of breeding experiments performed in the early part of the 20th century.

*I*N 1921 HEINZ HECK, FOUNDER OF Munich Zoo, began a series of breeding experiments in an attempt to recreate two extinct mammals — the European wild ox (aurochs), which died out in 1627, and the European wild horse (tarpan), extinct since 1887. The aurochs stood 6 feet high at the shoulder and had spreading, forward-curved horns. The tarpan was a small, dun-colored, ponylike horse with a short, erect mane and flowing tail. Heck believed that the genetic constitution of these extinct creatures still existed in the genes of various breeds of modern cattle and horses respectively, and that if these scattered features could be recombined by selective breeding experiments, the result would be a true aurochs or tarpan, recreated from their own descendants.

Tourist attraction
Although once extinct, the ponylike tarpan can now be seen in zoos throughout the United States and Europe.

Cattle crossing

Heck's experiments began with the cattle. He crossed several strains of cattle, each possessing features of the aurochs. Just eleven years after he began work, two calves were born that matched the descriptions, drawings, and old cave paintings of aurochs. Since then, more "aurochs" have been born and these still survive — apparent proof that some long-extinct species can be recreated.

Heck then tried to recreate the tarpan. He crossed various breeds of primitive pony and a breed of wild horse known as Przewalski's Mongolian horse, which possessed the erect mane typical of the tarpan. Once again his work was successful — reviving the hard-hoofed, broad-browed, dun-coloured tarpan with its short, erect mane and prominent eyes. Yet, even though Heck's modern-day aurochs and tarpans look exactly the same as their extinct ancestors, they may not be genetically identical.

The semistriped zebra

A similar project was recently initiated in South Africa, under the guidance of Reinholt Rau of the South African Museum in Cape Town. The project's aim was

Perfect replicas
The modern "aurochs" at Munich Zoo show the long, curved horns that were characteristic of the original creature.

to recreate the quagga, a semistriped, zebralike animal that died out in 1883. The quagga was traditionally classed as a separate species from the zebra, but in the 1980's scientists managed to extract quagga DNA from a preserved museum specimen. Painstaking research then proved that the quagga was a subspecies of the plains zebra *Equus burchelli*. Rau suggested that this indicated that the genes responsible for the quagga's semistriped coat had not been lost but had been dispersed among the plains zebra population. In 1987 he began crossing plains zebras, and hoped to recreate a quagga by the end of the century.

Modern-day dinosaurs?

Perhaps the most extra-ordinary theory to emerge from these "resurrection experiments" is the highly speculative idea that one day even dinosaurs may be restored to life. It is well known that fossilized pieces of amber (the resin of ancient trees) often contain well-preserved specimens of prehistoric insects that are many millions of years old. The insects became trapped in the sticky substance as it dripped down trunks and stems. In an interview in the magazine *International Wildlife* in 1987, entomologist Dr. George O. Poinar, of the University of California at Berkeley, suggested that one day a piece of amber containing an insect that had been entombed shortly after sucking blood from a dinosaur may be found. Poinar believes that if the DNA within the ingested blood cells could be extracted and cloned in the laboratory, this procedure might, in theory at least, be the first step to recreating a dinosaur.

IS THE DODO REALLY DEAD?

Do officially extinct passenger pigeons and Carolina parakeets still fly over remote parts of the United States? And have Steller's seacow of the Arctic, the Mauritius dodo, and Africa's zebralike quagga also survived?

THE DODO, A LARGE FLIGHTLESS PIGEON from the island of Mauritius in the Indian Ocean, was hunted to extinction by 1681, according to official history. But it may not be extinct. In *Secret Africa* (1936) the explorer Lawrence Green wrote: "Is the dodo dead? I have looked upon the bones of the dodo in the museum at Port Louis, Mauritius, and I have heard the island people talk of the bird. Now these people living in the very home of the dodo never speak of it as extinct. They will tell you that the dodo has been seen many times since 1681....They declare...that the dodo still lives in remote parts of the island, in inaccessible cliff caves and mountain forests."

It is unlikely, but not impossible, that dodos could survive today. Too many supposedly long-extinct species have been rediscovered for such a possibility to be dismissed automatically — especially when relatively inaccessible habitats still await exploration in Mauritius and the many tiny uninhabited islets nearby.

IS THE QUAGGA EXTINCT?

The quagga was an odd-looking relative of the zebra that had stripes only on the front half of its body. Herds of quaggas once roamed the plains of South Africa's Cape Province and Orange Free State, but by the 1880's they had been hunted into extinction. The last specimen died in the Amsterdam Zoo on August 12, 1883 — or so everyone thought.

But herds of brown, incompletely striped zebras have sometimes been seen since then in what is now Namibia. These sightings have not been verified, however. Skeptics suspect that these animals are not quaggas but a rare subspecies of zebra called Hartmann's mountain zebra that looks similar when the animals are shrouded in the African heat haze. In these conditions their stripes seem to blend together.

Sole survivor

Although the passenger pigeon was the most numerous species of wild bird ever known, by the summer of 1914 only one specimen remained alive, a 29-year-old hen bird called Martha Washington, housed at the Cincinnati Zoo. When she died on September 1, 1914, the nation mourned, and the passenger pigeon was declared officially extinct. But was it? Since 1914 many sightings of passenger pigeons have been reported. Some were almost certainly mourning doves, birds that look similar to passenger pigeons, but other sightings are harder to explain away.

In September 1929 a professor at the University of Michigan, Philip Hadley, was hunting in a virtually uninhabited wilderness area of Michigan's northern peninsula when he caught a fleeting glimpse of a pigeonlike bird with a pointed tail. His companion, a Mr. Foard, got a much clearer look at the bird

Dish of the day
Even though its flesh was tough and unpleasant-tasting, the dodo was butchered, first by Portuguese sailors and later by Dutch settlers on the island of Mauritius. By 1681 the bird was extinct.

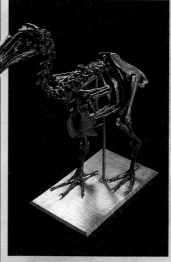

Dodo skeleton

Preserved specimen of the quagga

and had no doubt that it was a passenger pigeon. Foard was a very reliable eyewitness: as a young man he had seen the last of the great flocks of passenger pigeons winging across the skies. Alleged sightings of passenger pigeons have occurred more recently too. In March 1965 one such bird was allegedly spied by Irene Llewellyn at Homer, Michigan, and another was reportedly spotted during the same year at Park Ridge, New Jersey, by Stella Fenell.

The Carolina parakeet

The Carolina parakeet with its green plumage and yellow head was once very common in swamplands east of the Great Plains, until its liking for fruit made it a target for farmers and its pretty appearance led to trapping for the pet trade. It was believed that the Cincinnati Zoo owned the last Carolina parakeet, which died there in February 1918.

But since then, many yellow-headed green parakeets have been seen in its former haunts. In the 1930's three famous bird-watchers, ornithologist Alexander Sprunt, Jr., National Audubon Society president John Baker, and ornithologist Roger Tory Peterson, said they saw what may have been an immature specimen in the Santee swamp of South Carolina. Flocks of this species may still be hidden in the remote swamps of this area.

Steller's seacow

Steller's seacow — the largest modern-day member of the sirenian order of aquatic mammals, which includes the manatees and the dugong — was killed off only 27 years after it was first sighted. This gigantic creature measured 20 to 30 feet long and weighed up to $6\frac{1}{4}$ tons. It was discovered in 1741 and soon became a much sought-after source of meat in the Arctic. By 1768, the thousands of animals that had once existed had all been killed. Yet many huge sea mammals that resemble Steller's seacows have been reported since then.

In July 1962, for example, six strange animals were sighted in shallow water near the Kamchatka coast of Siberia by the crew of a whaling ship. The creatures were said to be 18 to 24 feet long, with dark skin, a small head with an abrupt transition to the body, an upper lip separated into two parts, and a sharply fringed tail. Scientists suspected that they might have been female narwhals (males have a distinctive long spiraled tusk). But the description, in the opinion of some experts, is more like that of a Steller's seacow than a female narwhal.

Passenger pigeon
Until they became extinct in 1914, passenger pigeons were so numerous that when they migrated, each flock contained several million birds.

> ## "Is the dodo dead? I have heard the island people talk of the bird. Now these people declare...that the dodo still lives in remote parts of the island, in inaccessible cliff caves and forests."
> **Lawrence Green**

The Tasmanian wolf
This skull is one of the few relics of this marsupial wolf, also called the thylacine. It allegedly died out in 1936, but it has reportedly been sighted in secluded regions of Tasmania, a large island off the south coast of Australia.

CHAPTER FIVE

THE WORLD OF PLANTS

Plants, sometimes considered the least exciting of the organisms in the natural world, are not just still, silent, and green. They display an amazing diversity and even have secret lives of their own.

The plant kingdom pulses with life, color, and movement. There are plants that trap and eat insects, shine beams of light, and mimic the smell of female moths or rotting flesh. From the realm of folklore come tales of plants that can devour human beings, flowers that can foretell the future, roots that let out unearthly shrieks when pulled from the ground, and fully formed live lambs born from enormous seedpods.

In conventional medicine as well as folklore, plants, when used sensibly, are known to have the power to heal. Plant extracts form the basis for many modern drugs, and new drugs are constantly being developed. For example, what may turn out to be a potential cure for the AIDS virus has recently been made from a little-known Australian chestnut tree.

One remarkable plant appears to be nearly human. For, like humans, it coughs and trembles when it breathes in dust. In 1900 the magazine *English Mechanic* published a report of a vine known as the coughing bean. A lover of moist tropical conditions, the coughing bean somehow spread to less favorable sites, particularly near railway tracks, which produced a strange result.

Violent trembling

The journal explained that the plant was tolerant of arid conditions, but: "there is one thing...it cannot stand and that is dust. When the breathing pores become choked by dust, the gases accumulate within the leaf for a time, and then are forcibly expelled in an audible paroxysm of coughing and sneezing which makes the leaf tremble violently." The plant even turns red just before it begins its coughing fit as the green chlorophyll particles sink down into the leaf and red particles rise to the plant's surface.

The fact that animals can move from place to place but plants cannot would seem to be one of the crucial differences between the two kingdoms, yet some plants are reportedly able to walk, albeit very slowly.

In India, an ancient mango tree has gained fame for its ability to walk. According to a 95-year-old villager, the tree has changed its location three times in the past 50 years. Dr. Ashok Marathe of Deccan College in Poona, India, says that the tree grows to a huge size,

Subterranean orchid
The Australian leafless orchid (Rhizanthella gardneri) *has the strange habit of living below ground. Its tiny purplish-red flowers develop under the earth, and only surface, after fertilization by burrowing insects, to allow the seeds to be distributed.*

then lowers one of its branches to the ground some distance away, where it takes root. Then the parent tree begins to wither and finally dies.

Speedy tendrils

Many other plants grow so quickly that they seem to move. One tropical climbing plant, *Cyclanthera pedata*, has the fastest growing tendrils in the plant world. It begins to curve its tendrils around a support within 20 seconds of first touching it. Only four minutes later the first coil is completed.

Other plants appear to move their leaves independently of the action of the wind. For example, the telegraph plant *(Desmodium gyrans)*, a native of Bengal in India, constantly waves its leaves in all directions, both up and down and even around in a circle, but why it does so defies explanation. It does seem to move more when in direct sunlight. And its cuttings continue to move for about 24 hours if they are kept in water.

Some plants are entirely guided by the sun, turning their leaves either to face its rays or to avoid its scorching heat. The plant *Silphium laciniatum*, found on the prairies from Texas in the south to Iowa in the north, is known as the compass

Supersensitive
The leaves of the mimosa plant curl up and droop when irritated. This double-exposure photograph shows one leaf cluster (below right) before and after being disturbed.

plant because of its ability to orient the edges of its leaves to face due north and due south. This ability is so marked that scientists at first thought its leaves might be magnetic. But no magnetic particles were found. Instead, it was discovered that the sun directs the orientation of the

> ## The plant's sudden movement may shake off insects which feed on its leaves, or by collapsing, it may make itself less visible to grazing animals.

leaves, whose surfaces are turned to face east and west. It is so accurate that when the pioneers were crossing the prairies they used the plant as a compass.

Touch-sensitive

The mimosa, or sensitive plant (*Mimosa pudica*), can also move its leaves. At the slightest touch, this shrub's leaflets fold up one after the other along the leaf stalks, at the rate of almost one inch per second; if given a hefty knock, leaves all over the plant may collapse as fast as 4 inches per second. About 10 or 15 minutes later, the leaves regain their normal upright positions.

Botanists now understand the mechanical process by which this happens: The folding of the leaves is caused by changes in water pressure in certain cells at the base of the leaflet stalks. When stimulated, the cells quickly lose a lot of water, which in turn causes the leaflets to collapse. But no one knows how the stimulus is transmitted through the plant or what real advantage the mimosa gains from being so very sensitive.

Lamblike fleece
The cotton plant, whose fluffy white boll is shown here, may be the origin of the legend of the vegetable lamb, an animal that is said to have sprung from a plant.

Changes in temperature can bring about the collapse, but other, equally delicate, plants survive meteorological shocks without using such a device. Some botanists have theorized that the plant's sudden movement may shake off insects which feed on its leaves, or that by collapsing, it may make itself less visible, or accessible, to grazing animals.

The art of self-defense

Most plants, however, cannot move in order to avoid predators. Instead they have developed an astonishing array of deterrents: from spiny leaves and stinging hairs to repulsive odors and poisonous juices. The most intriguing example of this self-defense comes from the veld of South Africa. There, antelope browse on the sparse, scrubby vegetation of acacias and other shrubs and trees. But they feed on any one plant for only a few minutes before moving on to the next. Why don't they carry on feeding until they have eaten all the leaves they can reach?

The answer, according to Lyall Watson in *Beyond Supernature* (1986), is simple: "The trees won't allow it. They put up their chemical defences — and they do so with astonishing speed." When an

Vegetable lamb

STRANGE FRUIT

Extraordinary plants with exotic fruits abound, but none are so strange as the vegetable lamb of Tartary, a legendary plant whose gourdlike fruit would apparently burst open to reveal a living lamb.

It became generally accepted that the legend had been inspired by the sight of the woolly root of a tree fern that had been carved in the shape of a lamb. The Royal Society of London received just such a carving in 1698.

Mistaken meaning

This explanation was accepted until 1887 when Henry Lee realized that the description of the vegetable lamb matched that of the cotton boll.

Peter Costello in *The Magic Zoo* (1979) finally solved the riddle: A Greek philosopher named Theophrastus had described wool-bearing trees that "bear no fruit, but the pod containing the wool is about the size of a spring apple." The ancient Greek word for apple also means "sheep"; so a "spring apple" could easily be mistranslated as a "spring lamb." The tree in question is the cotton plant, and its white boll, the vegetable lamb.

The dawn redwood

A SPECTACULAR FIND

In 1941, while studying the fossilized remains of conifer twigs and cones in Honshu, the Japanese botanist Miki observed that a few were noticeably different from the rest. Their leaves grew in pairs and their cones had long slender stalks, while the majority had short-stalked cones and leaves arranged alternately along the stem. The botanist was sure he had identified a whole new group of trees, and he named it *Metasequoia*, now popularly known as the dawn redwood.

A living fossil

Approximately 100 million years ago, the dawn redwood grew not only in eastern Asia but also in North America, northern Siberia, and Greenland. Twenty million years later, however, it apparently vanished from the face of the earth — until 1944 when a Chinese forester called T. Wang found a strange tree in Sichuan province, China, that he could not identify. Samples taken from it were shown by Chinese scientist H. H. Hu to be identical to fossil counterparts from the supposedly long-extinct *Metasequoia*. In 1948 more than 100 dawn redwoods were counted growing in the same area of China.

antelope begins to tear at the leaves of a plant, the plant releases a flood of chemical compounds, known as tannins, to its leaves, making them bitter and unpleasant to taste. The more leaves the antelope eats, the more unpalatable they will become.

This is the method the plant employs in self-defense, and it is extra-ordinary enough. What is even more remarkable is that the trees seem to be able to alert trees of their own kind in the vicinity to the danger that is threatening them. Researchers simulating the destructive behavior of the antelope found that trees nearby, which were not themselves under attack, showed a sympathetic increase in tannin levels. This may be the result of chemical substances known as pheromones being transmitted from tree to tree, but this is by no means certain. As Lyall Watson points out, if trees can warn one another of danger, they are little different biologically from animals that give alarm calls to warn their companions of the approach of predators. If plants do communicate with one another, they may be more like animals than we have so far believed.

Bamboo communication?

But some plants have an even more remarkable ability. They seem able to communicate not just with their close neighbors but with others of their species around the world. The flowering pattern of the bamboo, for example, seems little short of miraculous. Bamboos

A flowering Geiger counter?
The Japanese geneticist Sadao Ichikawa has produced strains of the spiderwort, a common roadside flower, that react strongly to doses of radiation. For unknown reasons, the hairs in the stamen turn from their natural blue to pink when exposed to radiation.

flower erratically, and at intervals that may be decades apart. Yet plants of the same species flower at the same time wherever they are in the world — with the exception of those varieties that grow near the equator. They do so even in very different climates.

Miraculous flowering

Stephen Jay Gould, professor of geology and zoology at Harvard University and author of *Ever Since Darwin* (1978), tells of one bamboo in Myanmar (formerly Burma) that was only 6 inches high,

---

Bamboos flower erratically, and at intervals that may be decades apart. Yet plants of the same species flower at the same time wherever they are in the world.

---

having been burned down repeatedly in jungle fires; yet even this plant flowered at the same time as its 40-foot-tall neighbors. It may be that the length of daylight somehow acts as a kind of long-term clock or calendar that enables them to count the passing years. The fact that the one place in the world where bamboos do not synchronize their flowering is within about 5° latitude of the equator, where the days and nights are always of equal length, lends weight to this theory. But how, or why, most bamboos synchronize their flowering across the world remains a mystery.

FABULOUS FUNGI

Living in a kind of twilight zone between the plant and animal kingdoms, fungi are some of the weirdest, yet most successful, living organisms on earth.

FUNGI ARE NOT REALLY PLANTS, due to the fact that they lack the vital ingredient called chlorophyll that enables green plants to utilize the energy of the sun. And even though their cell walls may contain chitin, the substance that many insects and crustaceans use to form their external skeletons, they are not classified as animals either. In fact they form a totally separate kingdom — and it is a very strange kingdom indeed.

Luminous fungi
These fungi are from the rain forest in Sumatra.

Strange fruit

The fruiting bodies of fungi come in various shapes — ranging from the familiar mushroom shape to shelf brackets, ears, and birds' nests. And they come in all sizes — from microscopic yeasts that are invisible to the human eye to giant puffballs that measure at least a foot in diameter (an American specimen had a five-foot circumference).

The largest fungus known to man is a giant tree-killing underground species called *Armillaria ostoyae*, which was discovered near Mount Adams in Washington. Terry Shaw, a scientist at the U.S. Forest Service Rocky Mountain Experimental Station in Fort Collins, Colorado, has estimated that it has been growing for between 500 and 1,000 years. It covers an area of 1,482 acres.

Silent hunters

It is through the mycelium, a mass of underground interwoven filaments, that the fungus feeds, absorbing nutrients from the plant or animal — whether dead or alive — on which it is growing. Some parasitic fungi produce a sort of glue that enables them to hold onto insect larvae while they penetrate the outer covering to get at the flesh. Other, microscopic species form their mycelium into something like a lasso with which they catch tiny eelworms that live in the soil; they then send out strands that penetrate the eelworm and digest and absorb its flesh.

When fungi invade plants they can be extremely damaging: At least 10 percent of the world's crops are destroyed each year by fungi, and entire forests have been wiped out by these silent killers. The fungus may break into the roots of the plant, gain entrance through wounds in the tree's trunk, or even invade the breathing pores (stomata) of the leaves.

But not all types of fungi are bent on destruction. Many form partnerships with trees and other plants that enable the host plant to survive. Some encourage their host's roots to grow and help them to absorb nutrients from the soil. They may even do battle with other fungi that threaten their partner, and prevent them taking hold. A few even feed their hosts.

A love-hate relationship

Orchid seed is so primitive that it needs an external source of food in order to germinate. A fungus known as *Rhizoctonia* usually provides this service. In return for a modest amount of food it takes for itself, it penetrates the seed and supplies it with nutrients. For a while the two organisms live in apparent harmony. But as the orchid grows and begins to make its own food through the photosynthesis process, it has no further need of the fungus and, in a fine display of ingratitude, it starts to digest the fungal filaments until the fungus is repelled.

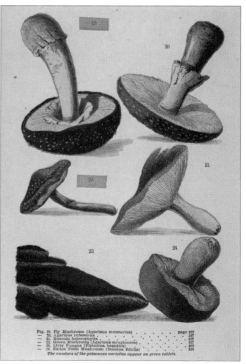

Fertile dust
At the slightest knock, as here from a raindrop, a ripe puffball will explode, releasing its contents in a cloud of dust, each speck of which could produce a new fungus. Some ripe puffballs can send out as many as 7 billion spores when they explode.

Delicious or deadly?
It is very difficult to tell with certainty an edible mushroom from a poisonous one.

PHARMACOPEIA OF THE FUTURE

For thousands of years plants have formed the basis of folk remedies. Western medicine is now beginning to realize the potential of these remedies and is using the plants on which they are based to develop new wonder drugs.

*D*URING THE YEARS OF PREHISTORY, humankind acquired a detailed knowledge of thousands of different plant species. This knowledge was passed on orally from generation to generation. The current use of medicinal plant remedies in traditional cultures throughout the world suggests that this faith in the healing power of plants is still strong. Over recent years, Western researchers have become increasingly interested in the use of plant ingredients. Some of the

Some of the world's largest drug companies are now sending out scientists to search for particular species as sources for drug products.

world's largest drug companies are now sending out scientists to search for particular species as natural sources for drug products. The gap between herbal and Western medicine could almost be said to be closing.

Tribal revelations
In Australia recent disclosures of some of the ancient secrets of Aboriginal folklore have yielded notable surprises about the medicinal potential of some of the country's plants. Tribal elders inhabiting the remote reaches of the Northern Territory have revealed that, when suffering from fever, the Aborigines of the region dry the leaves of the narrow-leaf fuchsia bush, boil them for two hours, then strain the resulting liquid and drink it. When they have backache, they take the inner part of a fan palm tree's stem base and pound it, soak it overnight, and then boil it for use as a poultice. And when their eyes become sore, they take two handfuls of leaves from the kangaroo bush, boil them, strain them, and use the resulting liquid as a soothing eye lotion. All of these natural remedies are remarkably effective, the Aborigines claim.

These, and many other traditional herbal remedies, were collected together during the late 1980's by an Australian research team and were documented in

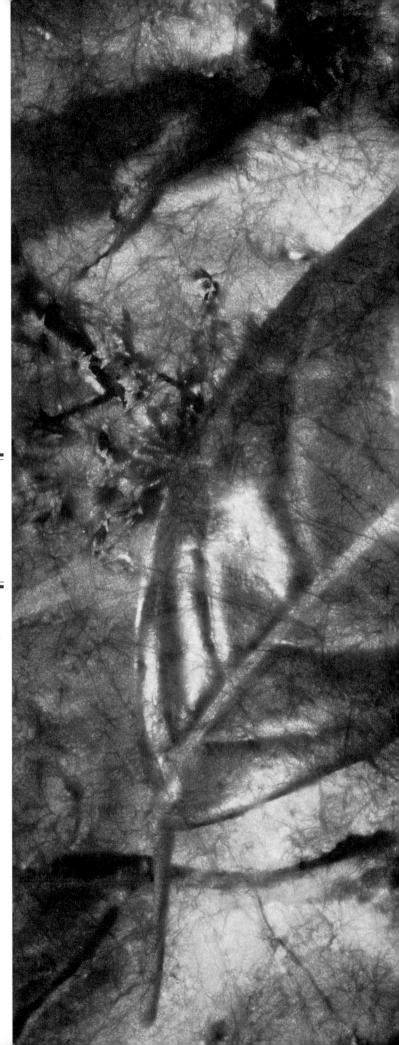

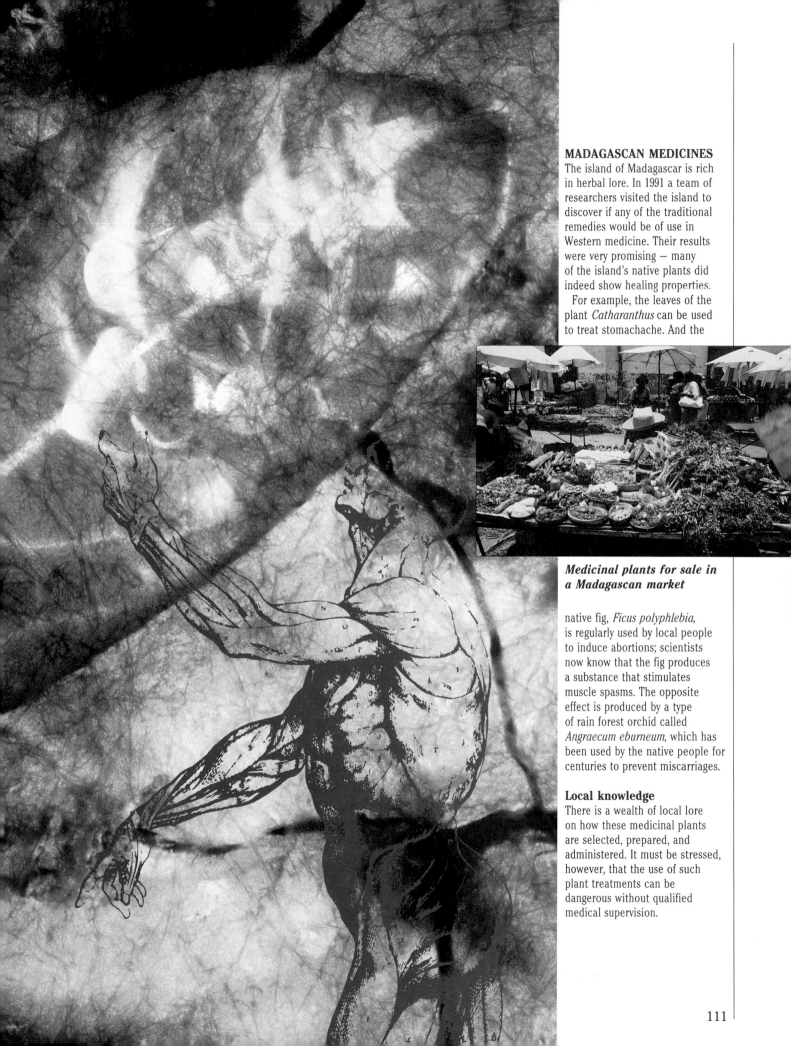

MADAGASCAN MEDICINES

The island of Madagascar is rich in herbal lore. In 1991 a team of researchers visited the island to discover if any of the traditional remedies would be of use in Western medicine. Their results were very promising — many of the island's native plants did indeed show healing properties.

For example, the leaves of the plant *Catharanthus* can be used to treat stomachache. And the

Medicinal plants for sale in a Madagascan market

native fig, *Ficus polyphlebia*, is regularly used by local people to induce abortions; scientists now know that the fig produces a substance that stimulates muscle spasms. The opposite effect is produced by a type of rain forest orchid called *Angraecum eburneum*, which has been used by the native people for centuries to prevent miscarriages.

Local knowledge

There is a wealth of local lore on how these medicinal plants are selected, prepared, and administered. It must be stressed, however, that the use of such plant treatments can be dangerous without qualified medical supervision.

a unique book, *Traditional Bush Medicines* (1988). And in 1987 an oil which was extracted from the Australian tea tree was marketed in Britain as an antiseptic and proved to be most potent. It had been found out that Aborigines have been using it as an antiseptic for thousands of years.

Indian cures

Herbal medicine has been a part of Indian tradition for centuries. In 1984 scientists discovered an Indian plant with the potential to control diabetes. Studies have revealed that the Indian scarlet gourd contains a compound that stimulates insulin release.

Currently, diabetics have to inject themselves regularly with insulin or take oral doses of insulin. If the compound in the Indian scarlet gourd could be isolated, it might be used to make drugs that stimulate insulin release within the body. Because this system would use the body's natural resources, it might help diabetics to lead healthier lives. Vital research is continuing in this area.

In the early 1990's Terence Moorhead, a British dentist working in India, was remarkably successful in treating his

> ## For hundreds of years, Indian herbalists have suggested that people suffering from stomach ulcers should eat plantains.

patients' gum disease when he began prescribing a toothpaste that contained compounds extracted from *Salvadora persica*, an Indian plant popularly termed the toothbrush tree, and used by the local people as a source of chewing sticks.

Another condition that may be ameliorated by an Indian folk remedy is stomach ulceration. For hundreds of years, Indian herbalists have

Plantain power
British researchers have recently shown that plantains contain a substance that is effective in treating stomach ulcers.

Natural antiseptic
Recent studies by Western researchers have shown that juice from the leaves of the Madagascan aloe seems to possess very effective antiseptic properties.

suggested that people suffering from stomach ulcers should eat plantains, a starchy relative of the banana. In 1988 researchers at Aston University in Birmingham, England, announced that they had identified the ingredient in plantains that was effective in combating such ulcers. The substance functions by stimulating the proliferation of protective cells in the lining of the stomach wall.

Herbal tranquilizer

Another plant drug that is now widely used is reserpine, which comes from the root of a plant called *Rauwolfia serpentaria*. The powdered root has been in use for more than 2,000 years in India for those suffering from mental illnesses. Not until 1952, however, when reserpine itself was isolated, did its use in Western medicine begin. Probably one of the most effective plant-based tranquilizers, reserpine was once widely used to treat schizophrenia. Today the drug is mainly used to control stress-related conditions such as high blood pressure.

During the 1980's, in the course of research into ancient Chinese folk cures, Chinese cardiologist Dr. Fan Lili visited

Massachusetts General Hospital, Boston, to investigate the alleged healing powers of a plant known to Chinese herbalists as Radix puerariae. In China, extracts of this plant are used by herbalists to treat headaches and high blood pressure. Dr. Fan discovered that the success of the plant was due to a compound contained in its roots that dilates blood vessels. What is so important about this discovery is the compound's potential as a drug for combating the effect of heart attacks. During a heart attack, the blood flow is drastically reduced, causing the death of organs or tissues as a result of oxygen starvation. By dilating the blood vessels, this compound might be able to allow oxygen to continue circulating.

Multiple healer

In 1987, anthropologist Kate Molesworth-Storer from Oxford University returned to England after a research trip to China's Xishuan Banna rain forest region with stories of a Chinese plant that appeared to cure a multitude of local ailments. Molesworth-Storer had learnt from the Dai people how they used a local wild plant, the bamboo orchid, to treat many types of illnesses, including sore throats, hepatitis A, mushroom poisoning, anemia, and even hangovers.

This natural multiple healer is being investigated by a team of British research scientists to discover if the claims are true and, if so, to pinpoint the active

ingredient responsible for its efficacy. Perhaps one of the most significant Chinese "rediscoveries," however, is the herbal drug called *qinghaosu*. Extracted from the shrub *Artemisia annua*, this natural remedy has been used for over 2,000 years by Chinese herbalists to treat malaria. It was not until 1991, however, that laboratory studies undertaken by the World Health Organization officially confirmed its success in the treatment of those diagnosed with malaria.

A cure for AIDS?

One of modern medicine's most urgent aims is to conquer AIDS. Some scientists believe that this conquest may soon be possible using nature's own pharmacy. Native to the subtropical and tropical forests of Queensland and New South Wales in Australia, the little-known Moreton Bay chestnut tree produces seeds that have been shown to contain a compound known as castanospermine. In nature, the compound acts as the tree's built-in survival system — it poisons any animals attempting to feed on its seeds and leaves.

But extensive laboratory tests carried out by Dr. William Haseltine at Harvard University, in Boston, revealed in 1987 that the compound had an exciting medical potential. It was found that castanospermine altered the sugars present on the surface of the virus that causes AIDS, thus preventing the virus from replicating and also from attaching itself to a host cell. Further detailed research is required to follow up this remarkable discovery, but it may offer some hope that a genuine cure for AIDS may indeed be within the grasp of medical science.

Dr. Linda Fellows

PLANT POTENTIAL
Dr. Linda Fellows, a plant biologist at Kew Gardens, in England, was the first person to bring to medical attention the potential of the substance

Black mulberry

castanospermine, produced by a little-known Australian chestnut tree, as an anti-AIDS drug.

In 1990 Dr. Fellows provided another breakthrough when she announced that she and a team of six scientists had uncovered a second natural compound, this time in the leaves of the black mulberry tree, that proved effective in the laboratory in slowing the progress of the HIV virus. Dr. Fellows has been awarded a grant by Britain's Medical Research Council to look specifically for natural anti-AIDS compounds. Her search continues.

Oriental methods

A Chinese herbalist makes up traditional herbal remedies for a variety of different conditions. The Chinese have been using herbal remedies since at least 3000 B.C.

PLANTS THAT EAT FLESH

Living in various parts of the world are bizarre plants that appear to reverse the rules of nature. Horror stories about meat-eating plants have fascinated people since they were first described in the 19th century.

TRAVELERS RETURNING FROM AFRICA in the 19th century enthralled their audiences with stories of the weird and wonderful wildlife they had seen there. But few stories struck such fear in the hearts of those who heard them as the tales of the terrible man-eating tree of Madagascar. According to these tales, anyone foolhardy enough to rest in the shade of this dreaded tree would be swept up in its giant tendrils and crushed to death, their remains no more than an empty skull tossed onto the ground beneath its awful canopy.

It is extremely doubtful that such a monster ever existed, since conclusive proof of its existence has never been found. But this is not the only account of a monstrous, carnivorous plant; other travelers told

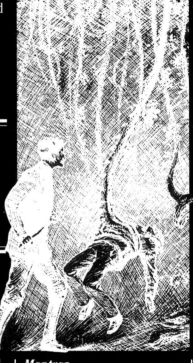

Anyone foolhardy enough to rest in the shade of this dreaded tree would be swept up in its giant tendrils and crushed to death.

stories of huge plants that could devour animals as large and ferocious as crocodiles. The truth, however, is not quite as dramatic. While some plants do eat flesh, the creatures that they are able to digest are only very small. In fact these plants feed on insects and small reptiles, not humans or crocodiles. Yet even if they are not the sinister man-eaters of local legends, the means they have devised to ensnare their prey are macabre enough.

Mantrap
This fanciful illustration by Fred White of a man-eating plant claiming a victim appeared in Strand *magazine in 1899.*

Engulfing leaves

Butterworts and sundews are two totally different types of flesh-eating plants that use similar means to devour their quarry. When an insect lands on the leaves of the butterwort, it quickly becomes trapped in the sticky mucus secreted by glands in the leaf. The leaf then curls inward to engulf the prey, and digests its meal using an acidic fluid. Sundews, which are common in Australia, also have leaves that curl to engulf their prey. The tentacles of sundew leaves end in a large droplet of fluid containing a slimy, sticky substance. Any insect that lands on the leaf is trapped by the slime. Then the tentacles move into action, each one bending over slowly. Eventually the whole leaf may roll up and completely surround the insect, so that it

No escape
For this hover fly, firmly stuck to the sticky tentacles of a sundew plant, death is imminent.

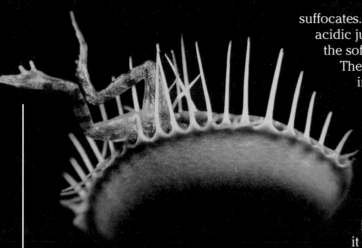

suffocates. Over about two days, the acidic juice in each tentacle digests the soft part of the insect's body. The leaf then unrolls and the indigestible parts of the insect are blown away.

Pitcher plants

Pitcher plants have leaves that make funnel-shaped traps to capture their prey. The rim of the trap is baited with nectar. When an insect lands in search of food, it loses its grip on the smooth waxy surface of the leaf and slides down into the trap. Hairs that point downward prevent the insect from crawling out, making death inevitable. Some pitcher plants secrete enzyme-rich juices that digest the prey. Others simply become filled with rainwater that drowns the insect, which will then decay. Special cells enable the pitcher plant to absorb the nutritious soup that results.

A fatal leap
A young frog is trapped in the interleaved spikes on the twin lobes of the Venus's-flytrap.

Funnel of death
Insects are attracted to the pitcher plant, shown here in a 19th-century illustration, by its smell and color. Once ensnared in the funnel-shaped trap, the prey cannot escape.

Prey trapper

The Venus's-flytrap is probably the best known of all the carnivorous plants. Each leaf of this ornate plant ends in a pair of rounded lobes, hinged at the center and edged with spine-like teeth. Three tiny hairs sit on the inner surface of each lobe; these form part of an intricate timing device, causing the lobes to snap shut as soon as the hairs are touched. But the lobes snap shut only when the object on the leaf is moving. This prevents the lobes closing unnecessarily when, for example, a raindrop or a leaf touches the hairs. Once the living prey is enclosed, the plant's digestive juices are set to work.

Carnivorous plants around the world have one thing in common. They all live in soil or water that is poor in nitrogen, a nutrient vital for growth. The flesh they consume helps to make up this deficiency.

Although scientists have yet to find any giant flesh-eating plants, there is, theoretically at least, no reason why carnivorous plants might not grow to huge dimensions and be capable of eating large animals or even humans. Perhaps one day, some intrepid — and extremely unfortunate — explorer may yet stumble on one of these man-eating vegetables.

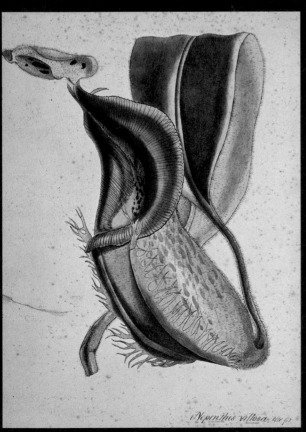

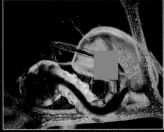

Bladderwort with prospective prey — a mosquito larva

A WATERY GRAVE

The little bladderwort has one of the most ingenious methods of catching its prey of any flesh-eating plant. The traps of these mostly aquatic plants operate with alarming speed at the slightest touch. The traps consist of very small bladders tha▮▮▮▮ed on delicate stems ▮▮▮▮▮ves.

Tiny triggers

Each of the bladders has tiny trigger hairs at the entrance. When the victim — usually a minuscule water flea — brushes against the hairs, the trap springs open and the flea is swept into the bladder with the water. Inside, the creature perishes through suffocation or lack of food, and its decaying remains are absorbed by special cells on the bladder wall. Once the creature's remains have been absorbed, the bladder opens again and the excess water is expelled. The bladder then returns to its original state and waits to be triggered again.

DEC

Some fl
that the
looking

M

could be
some plan
of enticing
  Rafflesi
vegetative
largest in t
jungle in S
grape vine
With a gig
covered w
looks like
it gives off
few days t

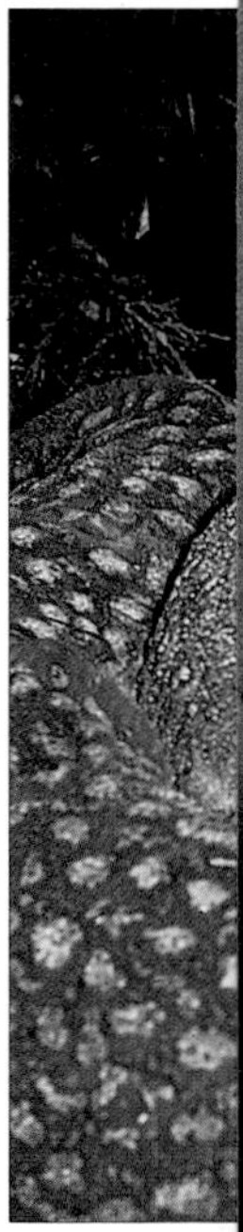

The world'
The flowers
can grow to

carrion fli
the secret
pick up po
to feed fro
next Raffle
  The gia
uses its un
it. The sm

MYSTERIES OF
PLANT BEHAVIOR

Many people are convinced that plants are sensitive to the thoughts and words of those who care for them, and that talking to plants will help them thrive. But can science ever conclusively prove or disprove this theory?

ONE DAY IN 1966, the polygraph expert Cleve Backster wired part of a polygraph, more usually known as a lie-detector machine, to one of his plants to see how it would react to the pouring of water onto its roots. He had expected his equipment to indicate a gradual increase in electrical conductivity as the water rose in the plant, but the polygraph recorder showed a steady decrease. He then wondered how the plant would react to a physical threat, and decided to burn the leaf to which the polygraph was attached.

Dramatic reaction

Even before he had moved to fetch the matches, the polygraph chart showed a dramatic upward surge, and it did so again when Backster came back into the room. When he *pretended* that he was about to burn the leaf, the plant showed no reaction at all. It was, so Backster says, as if the plant could differentiate between real and pretended intent — as if it could read his mind.

This was Cleve Backster's first attempt at using the polygraph machine to examine the reactions of plants. During the late 1960's, he carried out many similar

> **Even before he had moved to fetch the matches, the polygraph chart showed a dramatic upward surge. It was, so Backster says, as if the plant could read his mind.**

experiments on plants that left him in no doubt that he could communicate with them. With the help of other researchers, and using a variety of equipment, Backster went on to test various different plants — all of which, according to Baxter, seemed much more aware than scientists had previously believed possible.

The polygraph also reportedly recorded a reaction from the plants when other creatures nearby were in danger. Backster believed that they could even anticipate the movements of a spider trying to evade capture by a researcher. When some of his experiments were observed by a physiologist from Canada, however,

117

THE FIN
In the lat
windswep
northeast
grew alm
to overflo
barren so
Wheat gr
reached 4
plants gre
their norn

But no i
scientific
here. The
the Findh
their succ
communi
spirits wh
all around
to the pla
and how i

Paradise
Although
hippies an
were draw
is still the
reach suc
some reas
the origina
have left,

Working
Gardeners
Communit
plants in t

HERBAL TRADITIONS:
PLANTS THAT HEAL

During the days when witches and evil spirits were thought to walk the earth, ordinary mortals often turned to plants for protection.

*D*IFFERENT PEOPLES BELIEVED that different herbs could protect health and ward off evil. For the ancient Greeks, rue was the most potent. They made an oil of rue juice and the dewdrops from moonwort, with which they anointed the heads of those in need of protection. Among the ancient Egyptians the lotus flower was believed to lend its protection to all who inhaled its potent scent.

Strange as these ideas may seem, they often had some basis in reality. Malevolent spirits were thought to cause disease, and it was the job of the herbalist to drive those spirits out.

Medicinal qualities
But the successes they achieved may have been due as much to the medicinal qualities of the plants as to the faith of the patient. Plants contain a variety of essential oils, minerals, alkaloids, and other compounds that make them invaluable as antiseptics, antibiotics, stimulants, laxatives, anti-inflammatory agents, and anesthetics. But plants also contain chemicals that can provoke serious — and even fatal — reactions in some people. Before using plants or herbs in any "medicinal" way, it is vital to consult a qualified doctor.

St. John's wort
Legend has it that evil spirits fled from the very smell of this herb. In Germany it was sometimes worn as an amulet on St. John's Eve, when witches were believed to be out on evil errands. And on the Isle of Man in England, if anyone stepped on the plant after sunset it was claimed that a fairy horse would carry off the clumsy mortal for a wild night's ride.

Lotus flower
This illustration from an ancient Egyptian funerary papyrus shows the transformation of a man into a lotus. The lotus was called the sacred bean of Egypt, or, more poetically, the rose-lily of the Nile. The sanctity of the lotus was also celebrated in India and China, where it was seen as a symbol of fertility and purity.

Scarlet pimpernel
Few flowers were as effective at counteracting witches' spells as the scarlet pimpernel. According to legend, it sprang from the soil at Calvary when Christ's blood fell to the ground.

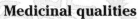

Mountain ash

English tradition has it that no witch will survive where the mountain ash, or rowan tree, flourishes. Using rowan wood for the rockers of a child's cradle was believed to ensure that the baby would be kept safe from evil.

Mandrake

Such was the power of the mandrake root that those who possessed it were believed to be able to heal the sick, put people to sleep, overcome infertility, and even drive out demons. Its magical qualities stemmed from the resemblance of the forked root to the human body. But such magic could be used for evil as well as good: While some believed that neither demons nor evil spirits could bear to look upon it, others said it carried the devil's very soul.

Garlic

Garlic was believed to have the power to ward off evil, offering protection from the evil eye, demons, witches, and vampires. Hung over a door, it was even believed to stop Satan from crossing the threshold.

Angelica

Every part of this plant was believed to be effective against evil. The aromatic oil is still used in liqueurs and perfumes, while the stem can be candied and eaten as a sweetmeat.

Thorn apple
Thorn apple, or jimsonweed, was one of the crucial ingredients of the ointments used by European witches in the Middle Ages. Its evil associations are thought to be due to its hallucinogenic properties. In larger doses, it can render people unconscious, and may even be fatal.

PLANTS THAT HARM

Plants and herbs have been used for evil as well as good. They were believed to help witches and bad spirits to carry out their malevolent deeds.

URING THE MIDDLE AGES and even later, people attributed witchcraft and certain evil deeds to certain plants. This ancient association is evident today in a number of plant names such as "witches' broomstick" and "devil's milk."

Oddly, some of the most beautiful flowers have evil associations. These might have arisen as a result of their poisonous, and sometimes even deadly qualities. Belladonna, for example (also known as deadly nightshade), is one of the most poisonous plants known. It is particularly dangerous because its glistening black berries are very tempting to children. But it is also one of a group of hallucinogenic plants that is now known to produce the sensation of flight. Witches were thought to have used such plants in their potions, which may explain why witches are typically portrayed flying through the night sky.

Unmasking evil
Some of the plants that were believed to possess supernatural powers are neither poisonous, nor used in evil potions. These plants were used in trials and judgments since they were supposed to be able to uncover the identity of witches and evil sorcerers who were pretending to be ordinary mortals. For example, dropping a walnut into the lap of a suspected witch would, it was claimed, prevent her from getting up, thus revealing her true identity.

Wolfsbane
This highly poisonous plant (also known as monkshood) was said to have sprung from the saliva of Cerberus, the three-headed hound who was believed to guard the entrance to the underworld by the ancient Greeks. Some medieval witches believed that applying an ointment made from the wolfsbane plant changed them into wolves. Scientists now know that ingredients in the plant irritate the nerve endings in the skin, causing a feeling of hairs rising.

Deadly nightshade
Belladonna is one of the most poisonous plants known. In the Middle Ages, this plant was thought to belong to the devil. It was believed that he guarded it so closely that he left it only on the night of April 30, to celebrate the Witches' Sabbath. Anybody who gathered the plant then was believed to be held firmly in the devil's power.

Vervain
According to legend, vervain was one of the most frequently used plants in witches' cauldrons. It is now known to contain a powerful hallucinogenic compound, which may explain why it was believed to enable witches to "fly with the devil." It is harmful even in small doses.

Henbane
Henbane is another poisonous plant that has long been thought to possess supernatural powers. The dark center of the flower looks like an eye peering out from among the leaves. It was known as the devil's eye and was believed to have helped the devil carry out his evil deeds.

Rue
In medieval times, rue was used to identify witches. If a woman vehemently rejected a gift of rue, she had undoubtedly sold her soul to the devil. It is also poisonous.

"Poison tree"
This 15th-century woodcut shows men sleeping beneath the branches of the "poison tree." Commonly found in graveyards, the yew tree was believed to bring misfortune to anybody who slept or ate in the shade of its branches.

CHAPTER SIX

WONDERS OF THE PLANET

Despite the advances of science, there are a number of natural phenomena that defy explanation. From earthquake lights to ringing rocks, from singing sands to shining seas, the earth and the oceans guard their secrets.

Within the heart of the earth lies a core of molten rock, ready to spill out in the form of lava through weak points on the earth's crust. Atop many of these locations volcanic mountains rise; linked together they form chains showing the places on the earth's surface where two continental plates meet and rub against each other. The movement of the plates causes earthquakes and volcanic activity. And they in turn cause unusual phenomena such as strange lights, weird noises, and changes in the earth's magnetic field.

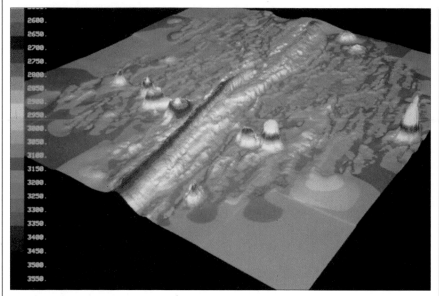

A mid-ocean ridge
This false-color three-dimensional map shows part of the East Pacific Rise, where two of the earth's plates meet beneath the ocean. Here magma wells up, cools, solidifies, and forms new igneous rock under the seas.

Where the tectonic plates meet under the oceans, volcanic islands sometimes burst through the surface of the sea. And far below the waves, seismic activity creates a strange environment known as the vent world. The creatures who live in this eerie world are unique.

Volcanic activity

Many severe earthquakes are caused when the ragged edges of the plates temporarily lock together, generating tension until the continental plates grind past each another, causing a tremor. Where one plate is actually forced underneath the other, the molten rock, or magma, melts and floats back to the surface, fueling the volcanoes above. Meanwhile new rock is constantly being formed along the plate margins in the middle of the oceans. This rock forces the plates apart, keeping them in a constant state of flux and causing earthquakes and volcanic eruptions elsewhere in the world.

Before, during, and after earthquakes and volcanic eruptions, odd

Studying the fire
Volcanologists, like these two scientists working in Iceland, monitor volcanoes to try to predict when they will erupt.

magnetic, electrical, and even luminous phenomena often occur. In many cases, scientists and researchers cannot explain what causes them.

There are frequent reports of bad-smelling gases emanating from the earth around the time of an earthquake. Gas may also be emitted during a volcanic eruption. At times these gases can be deadly. In one tragic case in 1986, a massive eruption of carbon dioxide from the bed of the volcanic Lake Nyos in Cameroon flowed downhill. The invisible gaseous cloud asphyxiated some 3,000 people living nearby.

Magnetic power

Peculiar magnetic effects have also been noted during earthquakes. In 1964 a tidal wave following an earthquake virtually destroyed the town of Kodiak in Alaska. A magnetometer sited on high ground above the town survived the quake. It showed that there had been a change in the magnetic field more than an hour before the earthquake struck. Similar effects were discovered around the San Andreas fault in California in the 1970's. Unusually large magnetic fluctuations were also detected before the eruption of New Zealand's Mount Ruapehu in 1969.

These fluctuations might be caused by the flow of electrically charged fluids in a volcano. Or pressure on the rocks during the buildup to an earthquake or eruption could cause them to release and conduct electricity. It might also alter their magnetic field.

The luminosity that often occurs during volcanic eruptions and earthquakes may also be caused by these strange electrical and magnetic effects. Japanese scientists have been studying such lights intensively since the early 1930's. In a typical instance, the sky is lit up by flashes of light, like sheet lightning but longer-lasting and considerably less bright.

In the South Hyuga earthquake of 1931, many witnesses saw

blue beams of light radiating from a spot on the horizon. Photographic evidence gathered during the series of earthquakes at Matsushiro from 1965 to 1967 seemed to indicate that the light usually lasted between 10 seconds and 2 minutes and

---

Where the tectonic plates meet under the oceans, volcanic islands sometimes burst through the surface of the sea.

---

that the central luminous body was a hemisphere, with a diameter between 65 and 650 feet. But these lights rarely occurred at the earthquake's epicenter. They appeared more often at mountain summits and known rock fault lines.

Thundering volcanoes

On May 8, 1902, Mount Pelée on the island of Martinique erupted. A glowing cloud of gas rose up from the mountain and rolled down the hillside to the town of St. Pierre, killing 29,000 people.

During the explosion, violent lightning played around the volcano. But lightning was also seen flickering around another mountain summit 40 miles away, which was not erupting. It was accompanied by glowing globes that exploded and emitted bursts of lightning. Occasional beams of light, like searchlights, were reported shooting out from Mount Pelée toward the second peak. Meanwhile, the lightning there became more intense.

Although the eruption in Martinique occurred nearly a century ago, scientists are no closer to being able to predict exactly when a volcano will explode. They do know that because of the fault lines beneath the earth's crust, some parts of the world are more prone to earthquakes and volcanic eruptions than others. Using this knowledge, they can predict which volcanoes may be likely to erupt. But if loss of life is to be prevented, forecasts in the future need to be more accurate.

Signs that a volcanic eruption is likely are the gradual deformation of the volcano and small changes in its local gravitational field, caused by movements of magma. In spring 1991, when Mount Pinatubo in the Philippines showed signs of being about to erupt, 90,000 people were evacuated. But monitoring is not always so successful. In 1980 at Mount St. Helens in Washington, which was being closely watched, scientists at the site could not predict either the actual moment of eruption or the extent of damage. Some of the geologists studying the volcano were killed when it erupted.

Scientists are studying certain changes that are known to herald a quake, in the hope of being able to predict when they

The island of Surtsey is born

THE MAKING OF AN ISLAND

In November 1963 an island emerged from the waters off the southern coast of Iceland. While the top of the volcano was still below sea level, violent explosions occurred. The hot lava vaporized the seawater and made steam. Once the pile of debris expelled by the explosions pushed the top of the volcano out of the water, lava began to flow again, and the newborn island's surface grew.

When the lava cooled, scientists began to study the island. They found it measured about a mile square and rose 560 feet above the waves, and 950 feet above the ocean floor. It was named Surtsey, for Sutur, Iceland's god of fire. In its early days, the island lived up to its fiery name: A column of steam four miles high rose from the volcano, and showers of ash fell for miles around.

First flowering

This small plant grows in the volcanic soil of Surtsey. At first, the newborn island was devoid of life, but now plants have sprouted from wind-borne seeds, and those dropped by seabirds visiting the island.

Blocking the flow
In some slow eruptions, the lava flow can be diverted to save lives and property. On Sicily's Mount Etna in 1992, huge concrete blocks were dropped from helicopters into the lava flow, successfully changing its course and keeping it away from houses and villages.

ENDING EARTHQUAKES
It is known that the filling of dams and the injection of chemical waste deep underground have been responsible for significant earthquakes (for instance at Koyna Dam in India in 1967). Scientists now know how to avoid causing them, and hope that someday they will be able to prevent earthquakes occurring.

One proposed method of minimizing the severity of earthquakes is to "lubricate" the jammed fault lines so that the stress within the earth is released gradually rather than quickly and destructively. At the moment, however, the risks are too great and scientific understanding too limited for such a strategy to be attempted as a deliberate attempt to control the destructive forces of the earth.

will occur. Among them are changes in the earth's electromagnetic characteristics, the build-up of strain in rocks, minute changes in gravity readings, and small bulges forming in the landscape. In addition, animals are often reported to behave strangely before the onset of an earthquake. These effects are being investigated in the hope that scientists will be able to predict the occurrence of both earthquakes and volcanic eruptions in the future.

Rocks are normally naturally silent. But some rocks or boulders (although seemingly the same as their neighbors) possess the remarkable ability to ring like bells when struck with another stone.

Three rocks or blocks of stone at Le Guildo, at the mouth of the Arguenon River in Brittany, France, give off a bell-like sound when struck. These rocks are made of a stone called amphibole, which is composed of silicates of manganese, calcium, and iron. The central rock, 20 feet long and 23 feet around, emits the clearest sound.

Ringing tone
What is it about this rock that gives it unique musical abilities? For one thing, it rests on a few pebbles and is thus slightly raised above the ground. This position seems to give its ring a remarkably clear tone. The two other rocks are partially sunk in the ground and emit only a muffled sound.

Other boulders in the cove also ring. The local people are able to recognize the musical rocks by their color — they are silver gray as opposed to the dark brownish color of their silent neighbors. These rocks also have a finer texture. An 1889 report in the magazine the *English Mechanic* claimed that there was, at the end of the cove: "a horizontal stratum partially buried in the shore, divided

---
A person listening at one end of the huge rock can hear a ringing noise when the opposite face is struck with a rock fragment.
---

into fragments, forming...something like the gigantic keys of a prehistoric piano. Three of these stones gave clearly the perfect major chord." Legend has it that these rocks guard the treasury of Satan.

Several of the stones that make up the vast megalithic remains at nearby Carnac give off bell-like notes when struck. At the monolith at Locmariaquer nearby, a person listening at one end of the huge rock can hear a ringing noise when the opposite face is struck with a rock fragment. Bernard Fagg, of the Jos Museum in Nigeria, noticed these effects in the 1950's.

Outrunning disaster
This magazine illustration shows refugees fleeing the eruption of Martinique's Mount Pelée in 1902.

Sacred stones
Ringing rocks are found at sacred sites around the world. Fagg first studied the rock gongs that are often situated near cave paintings and religious shrines in Nigeria. Although the gongs and paintings were ancient, they were still being used in modern secret religious ceremonies, often in connection with tribal initiation rites. They are found at Mbar, Bokkos, Daffo, and Fobur, all in central Nigeria. These rocks were also linked to fertility rites. At Nok, in central Nigeria, just prior to the harvest of acha, the local cereal crop, grass seeds were carried to the complex of gongs and ground down on the rocks.

Monitoring the San Andreas fault
Laser light is bounced back from reflectors on hillsides around Parkfield, California. This monitoring system can pick up tiny tremors and so, perhaps, give early warning of an impending earthquake.

Here and elsewhere deep grooves have been worn into the very hard granite, apparently as a result of corn grinding. Numerous other cultures, including those of India, China, and Japan, have made similar uses of ringing rocks.

The rocks found at Ringing Rocks Park, three miles north of Pottstown, Pennsylvania, were studied intensively by two researchers, John Gibbons and Steven Schlossman, in the late 1960's. These rocks range beween 1 and 15 feet in diameter and lie piled in woodland clearings. They are composed of a dark volcanic rock known as diabase, and were formed approximately 180 million years ago.

Rock research

Gibbons and Schlossman examined these rocks and concluded that they rang because one mineral, pyroxene, changed into another, montmorillonite. The molecules of that mineral are a different size, but the other minerals in the rock prevented the rocks from changing shape, which resulted in tensions building up. These pressures made the rock resonate at levels that could be heard as ringing tones. Rocks that did not ring had, according to their theory, been changed by exposure to moisture, perhaps by being in the shade or by being removed from the site and placed on damp ground, thereby easing the tension that was making them ring.

Yet this theory does not explain why the ringing rocks continued to ring, with the same tone, even when broken down to fragments a couple of inches long.

Ringing rocks
The musical rocks in Ringing Rocks Park, Pennsylvania, lie in well-drained woodland clearings. Some scientists believe that one reason they ring is because they are in a dry environment. The rocks in the shade of the trees are covered with lichen and do not ring, perhaps because they are too damp.

Sand dune tunes
The Gobi Desert in central Asia is the home of some remarkable singing sands. When people slide down the sandy slopes, the rushing sands produce a loud noise that makes the air vibrate as though the string of an enormous musical instrument had been plucked.

> The sands give off "a vibrant booming so loud that I had to shout to be heard by my companion. The weird chorus went on for more than five minutes continuously."
>
> R. A. Bagnold

Sand has also been known to make noise. Along certain stretches of the coastlines around the world, the sands emit sounds described as whistles or squeaks when put under sudden pressure, for instance by a heavy footfall.

Sand dunes and other sandy slopes either near the shore or in deserts have been known to make a loud booming sound. On the southwest coast of the Sinai Peninsula, Mount Abu Suweira or Gebel Naqus (which means "the Mountain of the Bell") has been known to the local people for centuries for its roar. Sand is forced up the incline of the slope by violent winds until it exceeds its angle of rest and can go no higher without sliding back down again.

In 1871 H. S. Palmer, a lieutenant in the Royal Engineers, noted the phenomenon in a survey report on the Sinai Peninsula. He said that after some small disturbance, the sand rolls "over the surface of the slope in thin waves an inch or two deep, just as oil or any thick liquid might roll over an inclined sheet of glass....The sound is difficult to describe exactly...sometimes it almost approaches to the roar of very distant thunder, and sometimes it resembles the deeper notes of a violoncello, or the hum of a humming top."

Deafening chorus

Some of these noises can be very loud and can carry long distances. One night the British adventurer R. A. Bagnold twice heard the sands give off "a vibrant booming so loud that I had to shout to be heard by my companion. Soon other sources, set going by the disturbance, joined their music to the first, with so close a note that a slow beat was clearly recognized. The weird chorus went on for more than five minutes continuously." The sound of the sands at Sand Mountain, a giant sand dune in Nevada, is so loud that it is claimed that a person standing nearby can be deafened. The noise can often be heard six or seven miles away.

Solving the riddle of the sands

What makes these sands so noisy? One theory is that the sand particles must be very pure and very similar in size for them to be able to vibrate in harmony and build up an amplifying resonance. The most widely accepted theories suggest that sand particles slide past one another in unison when forced to move, producing, in the case of singing sands, a soft sound. With booming slopes, however, the sound is produced by the whole moving mass of sand vibrating like

Sandy sounds
The sands of the Scottish island of Eigg sing when walked upon or touched. Scientists claim that it is the pockets of air around the quartz fragments making up the sands that cause this strange music.

a giant musical instrument. Perhaps the 19th-century British explorer Lord Curzon was right when he recommended that we "leave the Sounding Sands to continue their mysterious song, confining their favours to the lucky few, and exciting the curiosity, but, I hope, no longer the incredulity, of the remainder."

Genuine Sparklers

Diamonds are among the rarest and most beautiful of nature's creations. Over the years, many people have tried to cheat nature by attempting to make these gems in the laboratory. Yet can they succeed?

A real sparkler
Natural diamonds sparkle and glow with a distinctive fiery blue light.

ANCIENT LEGENDS CLAIMED that diamonds were created when dewdrops hardened. Others claimed that diamonds were alive. Hindu miners thought they grew like onions and that the larger they were, the older they were. Other miners thought that diamond beds renewed themselves, allowing a new crop to be harvested every 20 years.

Western scientists studied diamonds to see what they were made of. In 1673 the English physicist Robert Boyle heated a diamond: It disappeared in a puff of smoke. In 1772 the French chemist Antoine Lavoisier showed that diamonds broke down into carbon dioxide and carbon monoxide when burnt in air. Finally the English chemist Smithson Tennant showed that diamonds, despite all their mystique, are just a form of carbon, in principle not really very different from pure soot or graphite.

Making fakes

By the beginning of the 19th century, many people believed that diamonds could be created by applying extreme heat or pressure to carbonaceous materials, such as soot. Although these attempts to make diamonds failed, two of them did produce interesting results.

The Scottish chemist James Ballantyne Hannay mixed paraffin with hydrogen and the highly reactive metal lithium. The mixture was sealed in thick coils of wrought iron and placed in a furnace for several hours. The results were often dramatic — 77 out of 80 times, the tubes exploded, blowing up the furnace. But in three attempts, the tubes survived. Their interiors were coated with a hard, solid material in which a few transparent crystals were found.

These crystals were sent to M. N. H. Story-Maskelyne, keeper of minerals at the British Museum, who wrote to *The Times* of London on February 20, 1880, to say that there was no doubt that Hannay had created real diamonds. But others could not repeat the process. Yet in the 1940's X-ray diffraction tests seemed to prove that 11 of the 12 British Museum specimens did seem to

be diamonds. In the 1970's, scientists learned that fakes can usually be distinguished from real diamonds because the latter often glow with a blue light. Eleven of Hannay's diamonds had the characteristic blue glow of natural gemstones. Thus it seems that either Hannay's tubes were contaminated with diamonds, or he falsified his experiments, or he may indeed have discovered a way of synthesizing diamonds in such a manner that they showed a blue light.

Sparkling frauds

In the 1890's, the French physicist Henri Moissan, having noted that minute diamonds were found within iron-rich meteorites, tried to produce diamonds by plunging a mixture of molten carbon and iron (carbon will dissolve in iron) into water or molten lead. The combination of heat and pressure generated as the cooling iron contracted would, he hoped, create diamonds. When the iron was dissolved using acid, crystals were found that Moissan claimed were diamonds. Until recently his work had not been duplicated. Now, however, scientists have managed to produce crystals remarkably like those Moissan found. Yet these are either silicon carbide or alumna.

In 1953 the Swedish company ASEA proved that diamonds could indeed be created under high temperatures and at high pressures. ASEA and General Electric went on to develop commercial processes that typically operate at temperatures between 3,632°F (2,000°C) and 5,432°F (3,000°C) and pressures of over 100,000 atmospheres (some 1,500,000 pounds per square inch).

Interestingly, they found that to produce the fake gems, small diamonds had to be used as seeds. These seeds are then processed so that they grow slowly larger. Thus the miners who believed that diamonds could grow may not have been so wrong after all!

Diamonds under pressure
The comic strip hero Superman was able to make a diamond by crushing a piece of coal in his fist.

VOYAGE TO THE
SECRET VENT WORLD

Until recently, it was widely believed that only a few species of fish and other animals could live in the cold, dark ocean depths. But now exciting new discoveries of strange marine communities have dispelled that myth.

*I*N 1977, SCIENTISTS WERE AMAZED to discover the existence of unexpected concentrations of large and unusual animals, living among the hot-water springs that well up on the seabed at the boundaries between the tectonic plates that form the earth's crust. Initial discoveries near the Galápagos Islands, off Ecuador in the Pacific Ocean, were followed by finds in the East Pacific Rise south of the Gulf of California, and in the Marianas Trench in the west Pacific. Strange oases of life were found flourishing at depths of about 8,000 feet.

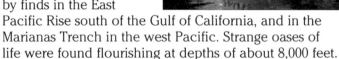

Yellow submarine
Scientists used this seven-foot-diameter submersible to explore a 4^1/$_2$-mile trench where the Nazca and Cocos tectonic plates separate from the Pacific plate. This trench houses 28 vent communities.

These unique collections of life-forms came to be known as vent communities, because of the mineral-rich, active, hot-water vents around which they live. Many of the species found are largely unrelated to other deep-sea creatures from the surrounding seabeds. These vent world species include giant worms, anemones, white crabs, lobsters, and grenadier fish, many of which have developed highly specialized forms to survive the high water pressure.

Feeding time
The strangest feature of these communities is their source of food. Other life on earth ultimately depends on light energy from the sun to build up organic matter, but vent animals depend on chemical energy derived from the vents. At the base of the food chain, chemosynthetic bacteria synthesize their organic matter by using chemical energy gained from oxidizing the hydrogen sulfide released by the underwater vents.

Some creatures feed on this profusion of bacteria in a conventional manner; others, such as the vestimentiferan rift worm, carry chemosynthetic bacteria with them in a special organ. These bacteria provide the worm with food and the worm provides the bacteria with a rich supply of oxygen. This animal has even done away with the need for normal feeding and, with it, the digestive tract.

Smoke in the water?
In the East Pacific Rise, these dramatic chimney-like volcanic vents, known colloquially as black smokers, billow out dense plumes of black, mineral-laden water.

Smoker crabs
This photograph, taken from the submersible *Nautile*, shows smoker crabs and the carnivorous, translucent zoarcid fish, amidst a tangle of Jericho worms.

Galápagos clams
Two-shelled mollusks up to 10 inches long are found in the Galápagos rift off the coast of Ecuador.

Curious cucumber
This giant sea cucumber was found on the seabed near the Galápagos Islands. It feeds on the sediment in the sea and can grow up to 18 inches in length.

Can of worms?
This collector can was used by scientists aboard the submarine *Alvin* to collect samples of marine life. Here, the can is used to scoop up samples of the giant rift worm, a species of pogonophoran worm.

Mysteries of the Deep

Across the surface of the sea, strange, luminous lights swirl and unexplained sounds echo, much to the bewilderment of people who see or hear them.

THE STEAMSHIP GLIDED THROUGH the still, tropical night in the Indian Ocean. The skies were clear and there was no moon. At 10:00 P.M. on that February night in 1880, the captain, R. E. Harris, was walking on deck before going to sleep when he noticed a peculiar white light on the horizon. The light seemed to be moving on a course that was converging with his vessel. If it continued, it would soon catch up with the ship. The object was unlike anything he had ever seen before, so he alerted the third officer.

A luminous circle

The crew came on deck to see the strange light. Later Harris wrote about it in the scientific journal *Nature*: "I was enabled,

> **"The ship was surrounded with one great body of undulating light. The waves stood above the water, like a highly luminous mist."**
>
> **Capt. R. E. Harris**

with my night glasses, to resolve in a measure what appeared, to the unassisted eye, a huge mass of nebulous matter. I distinctly saw spaces between what...appeared to be waves of light of great lustre. These came rolling on with ever-increasing rapidity till they reached the ship, and in a short time the ship was completely surrounded with one great body of undulating light....The waves stood many degrees above the water, like a highly luminous mist, and obscured by their intensity the distant horizon....Wave succeeded wave in rapid succession...."

Wheels around the world

The phosphorescent wheel that Captain Harris saw is like many others that have occasionally been reported, particularly from the Indian Ocean, over the centuries. Since the Second World War, people claim to have seen scores of them. Between October 1966 and March 1967 *New Scientist*, a British science journal, listed no fewer than five sightings between the northwest tip of Borneo and Bangkok. The wheels are usually said to be about a mile or more across, and to be made up of a number of spokes or spiral beams, rotating around a luminous hub. They may rotate clockwise or counter-clockwise; sometimes two wheels rotate in opposite directions at the same time.

Phosphorescence from marine organisms, perhaps stimulated by shock waves from a submarine earthquake, is the most popular explanation proposed for this phenomenon. But cases such as Captain Harris's, where the luminosity occurs above the surface of the water, cannot be explained in this way. Another theory is that this light is an electro-magnetic phenomenon that is caused by terrestrial discharges into the atmosphere and beyond, such as those seen during earthquakes and above mountain ranges.

Lethal tides

Sinister blood-red tides, which appear suddenly and leave a trail of death in their wake, have been known since time immemorial. Because no one knew what caused these tides, they were often attributed to divine anger. In the Bible, for instance, God tells Moses that he will help Moses turn the rivers of Egypt to blood so that the pharaoh will agree to release the Israelites. When Moses does so, the Bible reports that "the fish in the Nile died; and the Nile became foul so that the Egyptians could not drink water from the Nile." (Exodus 7:21).

Red tides are better understood now than they were in Moses' day. They are a type of algal bloom caused by tiny single-celled organisms such as dinoflagellates and diatoms and other photosynthetic microscopic algae.

One species responsible for these seas of blood is called *Noctiluca scintillans*. Its ancestors once grew like plants, but *Noctiluca* discovered that it could obtain more food by becoming

Fish kill
Dead fish float in the water off the coast of Sarasota, Florida, killed by a red tide.

Common cockle
Although shellfish are immune to the toxins produced by red tides, these toxins accumulate in their bodies, causing illness or even death to people who eat them.

Red tide
A section of the Sea of Galilee in Israel photographed from a satellite. The colors have been artificially enhanced. The lighter red areas at the rim and in the center show the presence of an algal bloom.

a predator and eating other microscopic animals. It has developed primitive eyes to allow it to find food, and a tentacle for swimming. This tiny, luminous creature is responsible for much of the phosphorescence that is present at low levels all the time in some parts of the world, for example, in the Mediterranean.

Unexplained toxicity

Noctiluca gathers at the surface in vast numbers, releasing toxins and making the waters look like a sea of blood, at unpredictable intervals and for reasons that are not completely understood. Red tides on the east and west coasts of America, and around Europe and Japan are also caused by *Noctiluca* and its relatives. In the 1960's and 1970's in Japan's Seto Naikai Sea these tides devastated the fisheries until pollution from sewage (which was stimulating their growth) was brought under control.

Yet, beyond knowing that pollution, on which *Noctiluca* thrives, may increase their overall frequency, no one is able to

are nearing death. The toxins they produce may simply be waste products that happen to be poisonous to other animals living nearby.

Strange sounds from the seas

Strange lights and seas of blood are not the only unexplained phenomena of the seas. Unnatural noises also abound in various parts of the world. Sounds known as water guns were once heard off many coasts. In northern France and Belgium *mistpouffers* (meaning "fog dissipaters") were well known to sailors in the vicinity

---

Blood-red tides, which appear suddenly and leave a trail of death in their wake, were often attributed to divine anger.

---

predict exactly when or where they will occur. Nor are we sure why they are poisonous. It has been suggested that the toxins are there to deter predators, but if so, this seems a peculiarly inefficient survival tactic. For those species that are able to absorb the remains of dead sea creatures, it may be a way of procuring food, but some of the dinoflagellates release toxins only when they themselves

of Ostend and Boulogne. In the 1890's M. van der Broeck, conservator of the Museum of Natural History of Belgium, wrote that the detonations are "dull and distant, and are repeated a dozen times or more at irregular intervals. They are usually heard...when the sky is clear, and especially towards evening after a very hot day." This is typical of the booming noises heard from the seas worldwide.

INFRASONIC ANSWERS

Infrasound waves occur at about 15 cycles per second, far below what the human ear can hear. But it may not be below what humans can feel. Some experts now believe that infrasound waves may cause the nervous feeling people often experience before bad weather or the occurrence of earthquakes. Because infrasound has such a low frequency, it can travel for hundreds of miles.

Although we are not consciously aware of it, we are constantly being bombarded by ultrasound from storms and other natural phenomena taking place up to 2,000 miles away.

Writing in *Ultrasonics* (1969), R. W. B. Stephens of the Physics department at Imperial College, London, claims that ultrasound can cause very real effects on people. "Physiological effects include nausea, disequilibrium, disorientation, blurring of vision, lassitude," he says. Other scientists, however, point out that research into the effects of ultrasound is still inconclusive.

Ganges Delta
The Barisal guns of the Ganges Delta in Bangladesh are one of the world's most famous unexplained noises from the sea. They are described as sounding like the booming of a cannon or the firing of a musket. The Ganges Delta is rich in natural gas. Some believe that the sounds are caused by gas bubbles rising to the surface.

Some scientists now believe that these sounds are caused by the release of large amounts of natural gas from beneath the sea. The continental shelf is pockmarked with holes that may have been made by natural gas exploding through the seabed.

The source of the sounds

In other areas of the world, such as Italy and the Philippines, these maritime rumblings are connected with imminent changes in the weather. In the Philippines they are harbingers of typhoons, while in Italy they presage rain, wind, or heat. The source of these sounds remains elusive. A contributor to the journal *Science* in the 1930's, Albert G. Ingalls, wrote: "like the foot of the rainbow, they are always 'somewhere else' when the observer moves to the locality from which they first seemed to come."

Today any unusual noises are usually explained away as the sound of low-flying military aircraft, the rumble of army bombing practice far away along the coastline, the underwater throb of the engine of a large ship, or the boom from seismic surveying for gas and oil. Some of these sounds can carry for many miles through the water.

But even in this noisy environment, reports of water guns continue to be recorded. In 1978 and 1979 strange noises were heard along the eastern seaboard and elsewhere in the United States. Some people thought that these sounds came from sonic booms from the transatlantic flights of the supersonic Concorde. But William Donn, a geophysicist at the Lamont-Doherty Observatory who operated a series of pressure-recording devices along the East Coast, was emphatic that this could not be. Some of the recordings, he said, were too far south, and the noise signature was very different from recordings of Concorde's sonic boom.

In some parts of the world a strange singing sound can be heard from the sea. In 1860 *Scientific American* carried an odd report. It was claimed that at various inlets around the Gulf of Mexico a rower, lying still in the boat, and with all other sounds absent, could sometimes hear coming from beneath the boat strains that sounded like a distant Eolian harp. Similar sounds were heard along the coast of Sri Lanka and attributed to a shellfish that sings at night when the moon is full. Neither of these songs has been fully investigated by scientists or completely explained.

The Greytown noises

People in iron ships anchored off the port of Greytown (now San Juan del Norte, Nicaragua), on the San Juan River, which separates Nicaragua from Costa Rica, heard a sound at regular, predictable intervals. In 1870 a correspondent to *Nature*, Charles Dennehy, described it: "While at anchor...we hear, commencing with a marvellous punctuality at about midnight, a peculiar metallic vibratory sound, of sufficient loudness to awaken a great majority of the ship's crew....This sound would continue for about two hours but with one or two very short intervals." Although some fish, such as the trumpet fish, make noises, Dennehy could not believe that they were the cause of such regularly occurring sounds.

Similar noises were heard in Trinidad. They reportedly sounded like a thousand Eolian harps, then changed to the noise of Chinese gongs under water, and finished by resembling a hundred human voices singing out of time in deep bass. To this day this complex array of sounds has not been completely explained.

Sounds heard along the coast of Sri Lanka were attributed to a shellfish that sings at night when the moon is full.

Siren song
Medieval stone relief of a mermaid. They were once thought responsible for strange singing heard at sea.

Musical fish
The underwater world is not silent. Whales are known to make sounds, and so do certain fishes, such as this trumpet fish. But they are probably not the only causes of the many strange sounds heard at sea.

FISHING FOR FOSSILS

She found herself face to face with an astonishing fish. Encased in thick, armor-like scales, it was blue with whitish spots.

ON DECEMBER 22, 1938, Marjorie Courtenay-Latimer, curator of South Africa's East London Museum, made a discovery that rocked the international scientific community. While examining a newly caught consignment of sea fishes at the local docks, on the lookout for specimens for the museum's collections, she saw an unusual fin sticking out beneath a pile of sharks; it was unlike anything she had ever seen before in her career.

Once the sharks had been removed, she found herself face to face with an astonishing fish. Five feet long, encased in thick, armor-like scales, it was steely blue with whitish spots. But to the young curator, the most extraordinary feature was its lobed fins, which resembled stumpy legs. Its tail fin was equally strange because it was composed of three lobes — it was completely different from the tail fin of any existing fish known to science at that time.

Miss Courtenay-Latimer purchased the fish and took it back to the museum to be preserved, and then wrote to South Africa's most eminent ichthyologist, Prof. J. L. B. Smith of Rhodes University. She enclosed a sketch of the fish and asked what it was.

Professor Smith was very familiar with the fish, but only as a fossil! For the creature was a coelacanth, a member of an archaic group of primitive fishes known as the crossopterygians, believed by some experts to have given rise to the land-living vertebrates. It was the first modern specimen ever seen by scientists, who had assumed that it had died out more than 60 million years ago, along with the dinosaurs. Smith formally named this surviving species of coelacanth *Latimeria chalumnae* in honor of the woman who had found it.

A scientific sensation

The coelacanth captured headlines worldwide; but despite Smith's efforts no other specimen was found until 1952. This second specimen (a deformed fish that was initially thought to be a new species) was captured near Anjouan, one of the Comoro Islands, northwest of Madagascar — hundreds of miles away from the South African coast where the first specimen had been caught. Moreover, although many coelacanths have been captured or filmed since then, all have been from the vicinity of the Comoros — no other specimen has ever been found near South Africa. Sadly enough, since its initial discovery, the coelacanth has become even rarer, for its populations have been depleted by irresponsible souvenir hunters and collectors.

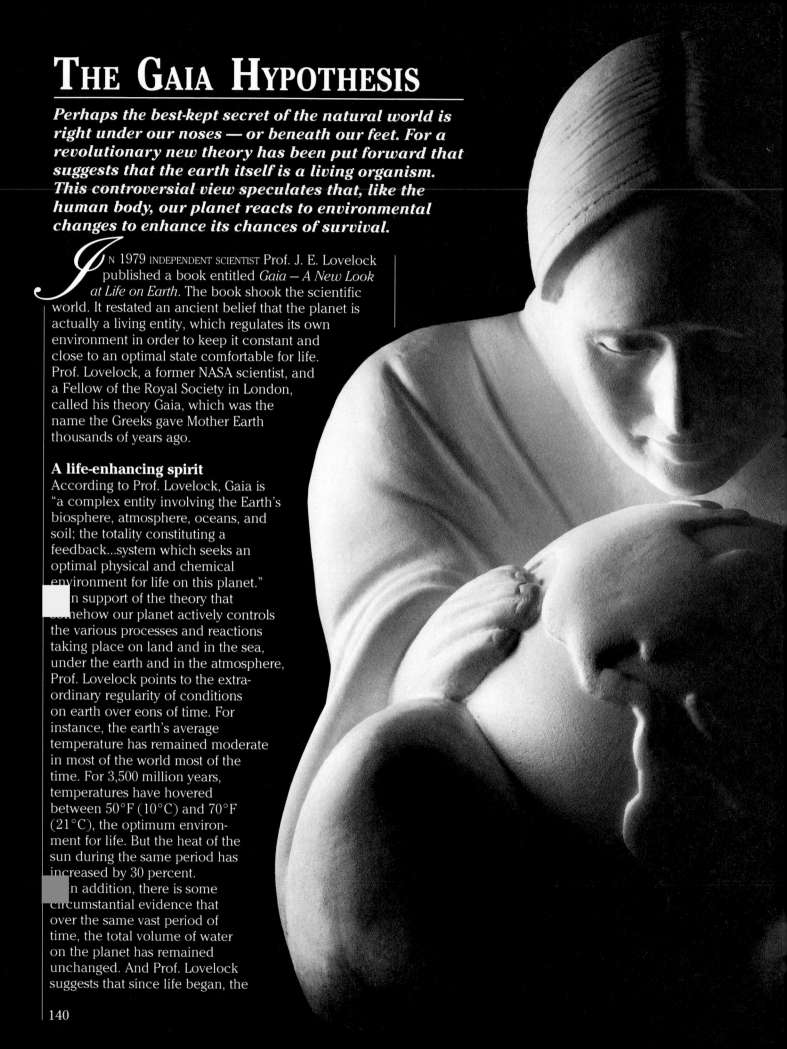

THE GAIA HYPOTHESIS

Perhaps the best-kept secret of the natural world is right under our noses — or beneath our feet. For a revolutionary new theory has been put forward that suggests that the earth itself is a living organism. This controversial view speculates that, like the human body, our planet reacts to environmental changes to enhance its chances of survival.

N 1979 INDEPENDENT SCIENTIST Prof. J. E. Lovelock published a book entitled *Gaia – A New Look at Life on Earth*. The book shook the scientific world. It restated an ancient belief that the planet is actually a living entity, which regulates its own environment in order to keep it constant and close to an optimal state comfortable for life. Prof. Lovelock, a former NASA scientist, and a Fellow of the Royal Society in London, called his theory Gaia, which was the name the Greeks gave Mother Earth thousands of years ago.

A life-enhancing spirit

According to Prof. Lovelock, Gaia is "a complex entity involving the Earth's biosphere, atmosphere, oceans, and soil; the totality constituting a feedback...system which seeks an optimal physical and chemical environment for life on this planet."

In support of the theory that somehow our planet actively controls the various processes and reactions taking place on land and in the sea, under the earth and in the atmosphere, Prof. Lovelock points to the extraordinary regularity of conditions on earth over eons of time. For instance, the earth's average temperature has remained moderate in most of the world most of the time. For 3,500 million years, temperatures have hovered between 50°F (10°C) and 70°F (21°C), the optimum environment for life. But the heat of the sun during the same period has increased by 30 percent.

In addition, there is some circumstantial evidence that over the same vast period of time, the total volume of water on the planet has remained unchanged. And Prof. Lovelock suggests that since life began, the salt content of the oceans has somehow been kept under biological control in order to provide the best possible conditions for life to continue.

Prof. Lovelock argues that the health of the planet would have been threatened from the beginning of time by disturbances in the environment caused by naturally occurring chemical reactions. The conversion of methane to carbon dioxide and of sulfides to sulfates, for example, would have shifted the balance of the biosphere to a more acid state, which life could not tolerate. But through a series of changes and processes that scientists do not yet completely understand, the earth has remained close to its present state of chemical neutrality as far back as scientific instruments can measure.

Earth's atmosphere contains approximately 21 percent oxygen. Twelve percent oxygen is needed to light a fire. But an increase of only 4 percent to 25 percent would be dangerous to life, for even damp vegetation burns at that percentage. If the atmosphere contained 25 percent oxygen, a small forest fire started by lightning might ravage a whole continent.

More than coincidence?

Prof. Lovelock argues that the methane and nitrous oxide released from the earth by natural processes regulate the oxygen content of the air. And the acidity of the biosphere is controlled by the production of 1,000 megatons of ammonia each year. This is precisely the amount needed to neutralize the sulfuric and nitric acids produced by natural oxidation of nitrogen and sulfur compounds. A naturally occurring coincidence, perhaps? But Prof. Lovelock thinks not.

Certain areas are crucial to the survival of the natural balance of the planet, in Prof. Lovelock's opinion. These areas, which include the tropical forests, scrublands, and the oceans' continental shelves, are

> In support of the idea that our planet actively controls the various processes and reactions taking place on land and in the sea, under the earth and in the atmosphere, Prof. Lovelock points to the extraordinary regularity of conditions on earth over eons of time.

where certain important regulatory chemical processes take place. Prof. Lovelock argues that concern over matters such as global pollution and the depletion of the ozone layer should take second place to the proper preservation of these special environments, for their survival is crucial to the survival of all life on earth.

Human guardians?

Prof. Lovelock's theories include humans in the ecological equation: He argues that humans, alone among the animals, have the capacity to collect, store, and process information, and then use it to manipulate the environment in a purposeful and anticipatory way. He even suggests that humankind, as one of the life-systems produced by Gaia, may possibly be intended for a role as the guardian of the planet.

INDEX

Page numbers in **bold** type refer to illustrations and captions.

A

Aborigines, 110, 112
Acacia tree, 107-108
Africa, 40, **40**, 100, 119
AIDS (acquired immunodeficiency syndrome), 106, 113
Alaska, 96, 126
Alaskan blackfish, 77
Albinos, **34**, **35**
Alexander, Charles, 58
Algae, 34, 47, 64, 136, **136-137**
Allosaurus (robot), **88**
Aloe plant, **112**
Alvarez, Dr. Walter, 90, **90**, 92
Alvarez, Prof. Luis, **90**, 92
Amber, 101
Ambun Stones, 39, **39**
Ammonites (marine family), 81, **82**, 85
Amphibians, 80, 81
Anacondas, 28-29, **28**
Anemones (vent community), 132
Angelica, **121**
Angraecum eburneum, 111
Animal Curiosities (W.S. Berridge), 31
Animals
    earthquakes, 52, 54-55
    psychic powers, 57-58
Animation, suspended, 64, **64**, 66-67
Anteaters, 67, 92
Antlions, 47
Aphids, 68
Arachnida, 99
Archaeopteryx, **92-93**
Arctic, 70, 77, **77**
Armadillos, 92
Armillaria
    mellea, 34
    ostoyae, 109
Artemisia annua, 113
ASEA, diamond experiments, 131
Attenborough, Sir David, 16
At the Earth's Core (movie), **89**
Aurochs, 101, **101**
Australia, 94, 100
    herbal remedies, 110, 112
    marsupials, **38-39**, 39
    pests, 71, **71**
    termites, **76**, 77

B

Backster, Cleve, 117-118, **118**
Bacteria, 47, 64, 80, 132
Badgers, 50
Bagnold, R.A., 130
Baker, John, 103
Bakker, Dr. Robert, 86
Bamboo, 108
Bänsiger, Dr. Hans, 41
Barisal guns, Ganges Delta, **137**
Baryonyx, 95, **95**
Bats, hibernation, **67**
Bears, 40, 67, 92
Becker, Josef, 57
Becquerel, Henri, 66
Beer, Trevor, 75
Bees, 50, 55, 73
    killer, **70**
    senses, 43, 45, 47, 51, **51**
Belladonna, 122, **122**
Benga, Zachious, 12-13
Beringe, Oscar von, 21
Berkland, Dr. James, **55**

Berridge, W.S. (*Animal Curiosities*), 31
Beyond Supernature (L. Watson), 107
Big cats, 23, 25, **25**, 40, 75, **75**.
    See also Cheetahs.
Birds, 41, 55, 77, 90, 93
    glowing, 33-34
    migration, 44-45, **44**, **45**, 46-47, **47**
    moon influence, 47
    singing, 31, 61, **61**
    vision, 51
Bladderwort, **115**
Bond, Henry, 24
Bornean cave racer, 31
Borneo, 31
Bottriell, Paul and Lena, 34-35
Boyle, Robert, 131
Brachiopods, 85
Branchiosaurs, **81**
Brazil, 71
    anacondas, 28-29, **28**
Breeding extinct animals, 101, **101**
Britain, big cats, 75
Brittany, France, 128
Broeck, M. van der, 137
Brontosaurus (Apatosaurus), 94
Brontotherium, 41
Buffaloes, **46**
Bull snakes, 30-31, **30**
Burgess Shale community, 83, **83**
Burma, snakes, 26
Burton, Dr. Maurice, 26
Butterflies, 35, **44**
    senses, **50**, 51
Butterworts, 114

C

Caldwell, Harry, 35
California, earthquakes, 54, **54**
Calyptra eustrigata, 41
Cambrian period, 80, 83, **83**
Camels, 92
Canada, 40, 83, **83**, 96
Carolina parakeets, 103
Carse, Mrs. Duncan, 31
Castanospermine, 113
Catfish, 54
Catharanthus, 111
Cats, 54, 55, 58, **58**, 75
    jungle, 75, **75**.
    See also Big cats; Cheetahs.
Cattle, 52, **55**
Chapin, Dr. James, 21-22
Chapman, Philip, 31
Cheetahs, 34-35, **34**
Chimpanzees, 61, **61**
China, 100, **120**, 129
    dinosaurs, 94, 95, **95**
    earthquakes, 52, 54
    herbal remedies, 112-113, **113**
Clams (vent community), **133**
Clark, Leonard (*The Rivers Ran East*), 40
Coccoliths, **87**
Cockles, common, **136**
Cockroaches, **66**
Codling moths, 71
Coelacanths, 139
Comoro Islands, 139
Compass plants, 106-107
Coral, 47, **47**, 80, 81
Corpse flower, 119
Costello, Peter, 107
Cotton plants, 107, **107**
Coughing beans (vine), 106
Courtenay-Latimer, Marjorie, 139
Crabs, 47, 132, **133**
    horseshoe, 47, 98, **98**, **99**
Creation, spontaneous, 35
Cresseveur, Crystal, 37
Cretacean period, 80, 81, 85-87, **87**, 89

Crichton, Michael (*Jurassic Park*), 88
Crocodiles, 61, 93
Crosse, Andrew, 35
Crowing crested cobras, 30
Crustaceans, 81, 93
Cryptobiosis, 66
Cuckoos, 46-47
Cyclanthera pedata, 106
Cyclops, 35, **35**
Cynognathus, **93**

D

Dai people, 113
Deadly nightshade (belladonna), 122, **122**
Deathshead hawk moths, **73**
Deinotheres, 96, 97
Dennehy, Charles, 138
Deserts, survival, 76-77, **76**, **77**
Dethier, Michel, 29
Dethier-Sakamoto, Ayako, 29
Devil-pigs, 39
Diabetes, 112
Diamonds, 131, **131**
Diapause, 76-77
Dilophosaurus, **91**
Dinosaur National Monument, **95**
Dinosaurs, 80, 81, **86**, **87**
    extinction, 85-87, 90-91, **90**, **91**
    fossils, 94-95, **94**, **95**
    movies, 88, **88**, **89**
Dodos, 102, **102**
Dog, Man's Best Friend, The (Capt. A.H. Trapman), 57-58
Dogs, **50**, 57-58, **58**, **74**
    earthquakes, 54, 55
Dolphins, 8, 11-14, 16, 47
    vocalization, 31, 60, **60**
Donn, William, 138
Dormice, 67
Dover cliffs, England, **87**
Ducks, 73

E

Earth, 80, 86, 140-141
Earthquakes, 126-127, 128, **129**
    animal behavior, 52, 54-55
East Pacific Rise, 126, **126**, 132, **132**
Egypt, lotus papyrus, **120**
Eigg, Scotland, **130**
Electricity, 55, 126-127, 128
Elephants, 31, 68
Ellenby, Prof. Conrad, 64
Ellis, William, 24
England, **87**, 95
Ethiopia, 68
Evans, Dr. Peter, 16
Evans, Norman, 75
Extinctions, 80-82, 85-86

F

Fabre, Jean Henri, 48
Fagg, Bernard, 128
Faraday, Michael, 35
Fawcett, Col. Percy, 28, **28**
Fellows, Dr. Linda, **113**
Fenell, Stella, 103
Ferns, fossils, **82**, **93**
Ficus polyphlebia, 111
Fikke, Quentin, 12
Findhorn community, **118**
Fireflies, 33

Fish, 33, 77, 81, 93, **138**
    Cambrian period, 80
    earthquakes, 54-55
    moon, 47
    red tides, 136, **136**
    vent communities, 132
Flatworms, 47
Flies, 65
Flynn, Rachel and June, 59
Foard, Mr. (passenger-pigeon sighting), 102-103
Folklore, 41, 74, 131
    Aboriginal, 110, 112
    plants, 105, 107, 110, 111, 114
Foraminifera (single-cell animals), 85
Forest Life and Adventures in the Malay Archipelago (E. Mjöberg), 38-39
Fossils, 79-81, **81**, **82**, 90, 96, 98, 99
    Cambrian period, 83, **83**
    dinosaurs, 94-95, **94**, **95**
    plants, **82**, 87, **93**
Foxes, 70, 92
Frogs, 55
Fujian province, China, 52, 54
Fungi, 34, 109, **109**

G

Gaia, 140-141
Gaia — A New Look at Life on Earth (Prof. J. E. Lovelock), 140-141
Galápagos Islands, 132, **133**
Galazzo, Vincent, 14
Garlic, **121**
Gastropods, 93
Gazelles, **21**, **46**
Geese, **45**, 46, 52
Genes, mutations, 34-35, **34**, **35**
Geology periods, 80-81, **80**
Germany, 90
Gertie the Dinosaur (movie), **89**
Gesner, Conrad (*Historia Animalium*), **37**
Gibbons, John, 129
Ginkgo biloba, 100
Glowworms, 33
Glyptodonts, **39**
Goats, forest, 23, **23**
Gobi Desert, 95, **130**
Goldfish, 54
Gorillas, 20-21, **20**, 35, 61
Gould, Stephen Jay, 108
Gourds, Indian scarlet, 112
Great Barrier Reef, 47
Greece, 65, 120
Green, Lawrence (*Secret Africa*), 102
Greenland, 96
Greenwell, J. Richard, 23
Grogan, Ewart, 20-21
Groves, Dr. Colin, 21
Grunion fish, 47
Gubbio, Italy, 90
Gulf of Mexico, 138
Gulls, 46, 59

H

Hadley, Philip, 102
Haicheng, China, 54
Hallucigenia, **83**
Hamsters, 67
Hannay, James Ballantyne, 131
Hares, snowshoe, 70, **70**
Harris, Capt. R.E., 135-136
Haseltine, Dr. William, 113
Heck, Heinz, 101
Hedgehogs, 67

Hemmer, Dr. Helmut, 23
Henbane, **123**
Hens, 61, 73
Herons, great blue, 34
Hessian flies, 71
Hibernation, 41, 66-67, **66**, **67**
Hinton, Prof. Howard, 76
Hippopotamus, pygmy, **21**
Historia Animalium (C. Gesner), **37**
Hogs, giant forest, **21**
Horner, John R., 95
Horses, 92
Horsetails *(Equisetum)*, **100**
Hu, H.H., 108
Hummingbirds, **44**, 51
Hyenas, 50

I _____

Iceland, volcanoes **126**, **127**
Ichikawa, Sadao, 108
Ichthyosaurs, 85, **90**
In Darkest Africa (H.M. Stanley), 20
India, 112, **120**, 129
Indian Ocean, 135-136
Ingalls, Albert G., 138
Insecticides, 65, **65**
Insects, 35, 80, 81, 101
    resistance to chemicals, 65, **65**
    temperature, 77
    U.S.A. pests, 71
    vision, 51, **51**
Iridium, 92
Irkuiems, 40
Italy, 65, 138
Ituri forest, Zaire, 20, 21-22
Ivory trade, **97**

J _____

Jaeger, Dr. Edmund C., 41
Jakun people, Malaysia, 41
Japan, 54-55, 129, 136-137
Jellyfish, 80, 83
Jerboas, **76**
Johnston, Sir Harry, 20
Jungle cats, 75, **75**
Jurassic Park (movie), 88

K _____

Kalahari Desert, 28
Kamchatka Peninsula, Siberia, 40, 103
Kellas cats, 75
Kerchelich, M.F., 26
Kerr, Prof. Warwick, 70
Kewalo Basin Marine Mammal
    Laboratory, 60
Klinowska, Dr. Margaret, 16
Knott, John, 30
Koala bears, 67, **67**
Kodiak, Alaska, 126
Koyna dam, India, 128
K-T event, 85-87, 90, **90**, 92-93
Kyoto University, Japan, **61**

L _____

Lake Nyos, Cameroon, 126
Lampshells *(Lingula)*, 98
Land that Time Forgot, The (movie),
    88, **88**

Lankester, Prof. Edwin Ray, 20
Lapwings, 47
Lavoisier, Antoine, 131
Lay, Dr. Douglas, 21
Lee, Henry, 107
Leeches, 55
Le Guildo, Brittany, 128
Lemmings, 70, **70**
Liaoning province, China, 52
Lice, **65**
Lierde, Col. Remy van, 28
Lights, phosphorescent, 135-136
Lili, Dr. Fan, 112-113
Limpets, 44, 47
Lingula, 98
Livingstone, David, 30
Lizards, aquatic, 85
Llewellyn, Irene, 103
Locusts, 68, **68-69**, 70
Long Marine Laboratory, 31
Lost World, The (movie), 88, **89**
Lotus, 120, **120**
Lovelock, Prof. J.E., 140-141
Luckett, Doreen, 74
Lungfish, 100, **100**

M _____

MacFarlane, Roderick, 40
Mackerel, horse, 73
MacKinnon, Dr. John, 23, **23**
Madagascar, herbal remedies, **111**
Magic Zoo, The (P. Costello), 107
Magnetism, 126-127, 128
    migration, 45, 47
Maguire, Adam, 8, 11
Malaria, 113
Malaysia, 41
Mammals, 47, 80, 87, 90, 93
    marsupials, 39, 85, 92
    placental, 92
    temperature, 77
Mammoths, 96, **96**, **97**
Mandrakes, **121**
Mantell, Walter, 22
Marathe, Dr. Ashok, 106
Marsh, Prof. Othniel C., 41
Marshall, Robert E. *(The Onza)*, 23
Marsupials, 39, 85, 92
Marx, Bob, 11
Mastodonts, 92, 96, **97**
Matsushiro, 126-127
Mauritius, 102
McAtee, Dr. W.L., 33-34
McKay, Winsor, 89
Medicines, herbal, 105-106, 110, **111**,
    112-113, **112**, 120, **120**, **121**
Menzies, J.I., 39
Mermaids, **138**
Merriam, Prof. C. Hart, 40
Messina, Sicily, 54
Metasequoia (dawn redwood), **108**
Mexico, 23
Mice, 31, **31**, 37, 77
Michigan, 103
Midges *(Polypedilum vanderplanki)*, 76
Migration, 41, 44-47, **44**, **45**, **46**, **47**
Miki, 108
Milne, A.A. *(Winnie-the-Pooh)*, 40
Mimosa plants *(M. pudica)*, **106**, 107
Minhocao, 39
Mjöberg, Eric, 38-39
Moissan, Henri, 131
Mole rats, 50, **77**
Molesworth-Storer, Kate, 113
Mollusks, 80, 85
Moloney, Jason, 8, 11
Monarch butterflies, **44**
Montana, 87, 91
Moon, influence, 47
Moonworts, 120

Moorhead, Terence, 112
More Animal Legends (M. Burton), 26
Moreton Bay chestnut, 106, 113
Mosasaurs (aquatic lizards), 85
Mosquitoes, 65, 76-77
Mosquito fish, **65**
Motherhood, 61
    fostering, **72**, 73, **74**
Moths, 41, 48
Mount Abu Suweira, 130
Mountain ashes (rowans), **121**
Mount Etna, Sicily, **128**
Mount Pelée, Martinique, 127, **128**
Mount Pinatubo, Philippines, 127
Mount Ruapehu, New Zealand, 126
Mount St. Helens, Washington, 127
Mulberry, black, **113**
Murillo, Andres Rodriguez, 23
Mushrooms, **109**
Mutations, genetic, 34-35, **34**, **35**
Mystery Cats of the World (Dr. K.P.N.
    Shuker), 75

N _____

Nathawat, J.S., 25
Naud, Marlène, **60**
Nautiloids, 81, **100**
Navajo Indians, 41
Nematodes, 64, **64**
Neoceratodus, 100, **100**
New South Wales, **82**
New Zealand, 16, 22-23, **22**, 100
Nigeria, 128
Nijnatten, P. van, 36
Nikita (pony/zebra), **74**
Noble, Edward, 75
Noctiluca scintilans, 136

O _____

Oglala Sioux Indians, 41
Okapi, 20, **20**
Oman, whales, 16
One Million Years B.C. (movie), 88,
    88-89
One of Our Dinosaurs is Missing
    (movie), 88
Onzas, 23, **23**
Ophrys, 119, **119**
Opossums, 92
Orangutans, 61
Orbell, Dr. Geoffrey, 23
Orchids, 113, **106**, 109, 119, **119**
Orodromeus, 94-95, **94**
Owls, 33, 70

P _____

Pacific Ocean trenches, 132, **132**, **133**
Palmer, H.S., 130
Palorchestids, **38-39**, 39
Pangaea, 81, **86**
Papua New Guinea, 39, **39**
Parakeets, Carolina, 103
Parasites, 74
Parrots, **60**, 61
Passenger pigeon, 102-103, **103**
Patagonia, 92
Payne, Katie, 31
Peacock moths, 48
Peacocks, **35**
    Congo *(Afroparva congensis)*,
        21-22, **21**
Permian extinction, 80-82, 86

Pets, 57-58, **58**, **60**
Pfefferkorn, Father Ignaz, 23
Pheromones, 48, 50
Philippines, 138
Phosphorescence, 33-34
Piggott, Sir Digby, 33
Pigs, 35, 52
Pikaia, **83**
Pitcher plants, 115, **115**
Plagues, 68, **69**, 70, 71
Plantains, 112, **112**
Plants, 80, 81, 90, 105-108, **106**, **107**,
    108
    carnivorous, 114-115, **114**, **115**
    flowering, 80, 91, **93**
    fossils, 87, 88
    hallucinogenic, 122, **122**, **123**
    medicinal, 105-106, 110, **111**,
        112-113, 120, **120**, **121**
    reactions, 117-118, **118**
Pleimes, Dr. Ute, 57
Plesiosaurs, 85, **92**
Pliny the Elder, 33
Poinar, Dr. George O., 101
Polar bears, 34, **74**
Polar regions, 70, 77, **77**
Ponies, Shetland, **74**
Poorwills, 41, **41**
Popilla japonica, 71
Populations, 68, **69**, 70, 71
Porcupines, 47, 92
Portuguese men-of-war, 73
Postkoks, 26
Pterosaurs, 85, 87
Puffballs, **109**
Pumas, 75, 92
Pythons, 29, **29**

Q _____

Qinghaosu, 113
Quaggas (zebralike animal), 101, 102,
    102

R _____

Rabbits, 50, 71
Raccoons, 35
Radiation, 82, 66, **108**
Radix puerariae, 113
Rafflesia, 119, **119**
Rahm, P.G., 66
Rats
    kings, 36-37, **36**, **37**
    earthquakes, 52
    kangaroo, 77
Rattus rattus, 36-37, **36**, **37**
Rau, Reinholt, 101
Raup, David, 93
Rauwolfia serpentaria, 112
Red Ghost, 71
Redwoods, dawn *(Metasequoia)*, 108
Remedies, herbal, 105-106, 110, **111**,
    112-113, **112**, 120, **120**, **121**
Reptiles, 80, 81, 100
Reserpine, 112
Rhinoceros, 50
Rhizanthella gardneri (orchid), **106**
Rhizoctonia (fungus), 109
Rhynchocephalia (prehistoric reptile
    group), 100
Ringing Rocks Park, Pennsylvania,
    129, **129**
Ritchie, Col. G.B., 30
Rocks, ringing, 128-129, **129**
Rowland, Albert, 25
Rue, 120, **123**
Rukwa Valley, Zambia, 35